Property and Money

Property and Money

Michael Brett

Cartoons by Nick Newman

A member of Reed Business Information

Estates Gazzette
151 Wardour Street, London, W1F 8BN

First published 1990
Second edition 1997
Reprinted 2001
Reprinted 2004
Reprinted 2006
Reprinted 2007

ISBN 0 7282 0278 6

Typesetting by Keyword Typesetting Services Ltd.
Printed in Malta by Progress Press Ltd

Acknowledgements

I am indebted to many people in the property world and the City of London for information and advice in connection with the series of articles on which this book is based. While it would be impossible to name them all, I would like to thank in particular Prudential Property Managers and Andrew Baum (then Property Research Manager at the Prudential, now Professor of Land Management at the University of Reading) for detailed advice on the early chapters.

Norman Bowie, consultant to Jones Lang Wootton, Gordon Pipe of Healey & Baker and Russell Schiller of Hillier Parker gave the benefit of their considerable experience on a range of topics, and Richard Mully of County NatWest and Susan Marks of Bank of America helped greatly with the sections on bank lending.

I am extremely grateful to Jack Rose – property developer, author and part-time academic – not only for help with the section on leasehold valuation but also for his efforts over more than two decades to din into me the arithmetic that lies behind property market phenomena.

Last but not means least, Piers Venmore Rowland of the City University's Department of Property Valuation and Management and Susan Marks very kindly read the manuscript of the book and made many invaluable comments for which I am extremely grateful. And without Alex Catalano of *Estates Gazette*, who had the original idea for the series of articles and contributed at every stage, this book would never have seen the light of day. The views expressed and the mistakes are my own responsibility.

July 1990

I have again had a great deal of help from the property industry and elsewhere in preparing the additional material incorporated in this revised edition. I would like to thank particularly Gerald Blundell of Jones Lang Wootton and Paul McNamara, property research manager of Prudential Portfolio Managers, for their comments on certain of the new chapters.

March 1997

Contents

continued overleaf

Contents

continued from overleaf

Introduction

This book is about the financial and investment aspects of commercial property. It is aimed not only at those in the property business — which includes students — but also at bankers, fund managers, accountants and solicitors whose work brings them into contact with commercial property transactions. The book attempts to bridge the comprehension gap between the financial markets of the City of London and the world of the surveyor, the estate agent and the property company.

It starts from first principles and does not assume existing knowledge. While the principles are timeless, the way they manifest themselves changes with economic conditions, with the tax regime and with investment fashion. So a little background is needed to the climate in which the book was written.

This, second, edition of "Property and Money" was prepared for publication in 1995-1997. At this point the commercial property industry had partially emerged from its worst recession in living memory, though conditions in some areas remained bleak. The period 1990 to 1993 had seen many of the industry's most cherished assumptions overturned. Rental levels slumped, investment yields rose and capital values of many types of property dropped like a stone. Property trading companies went bust right, left and centre and banking lenders wrote off billions of pounds in dud property loans. The new material incorporated in this expanded second edition analyses some of the effects and implications of this cataclysmic period in the property industry, while the original material has been updated and amended where necessary for changes since the first edition went to press in 1990.

The original edition of "Property and Money" was based on a series of weekly articles published in *Estates Gazette* in 1989 and early 1990. Much of the additional material in this new edition also appeared first in *Estates Gazette* in article form. When the first edition went to press the commercial property market was coming to the end of one of its periodic booms, when signs of overexpansion and coming financial problems were already looming. Over the previous few years, a number of important changes had taken place in the property scene.

First, Britain had enjoyed a development boom which had led to oversupply of space in some areas, notably offices in and around the City of London. It had been fuelled by rapid escalation in rents and property values. A long consumer boom was also coming to an end, raising anxieties as to the outlook for retail property, where rents had often doubled in little more than two years.

Second, the development boom had been very largely financed by the banks, whose loans to property companies were over £30bn by 1990 — they later went up to over £40bn. By contrast, the savings institutions — life assurance companies and pension funds — which had been the main long-term investors in commercial property in the 1970s and early 1980s, were playing a much less significant role. Replacing them in the market — for a time at least and for certain specific types of property — foreign buyers of UK property had come into prominence.

Property market observers worried whether tenants would be forthcoming for all the new space created during the boom and whether there would be long-term investment buyers for the developments financed short term by the banks. Such buyers were needed to allow the banks to get their money back.

Interest rates early in 1990 stood at very high levels, compounding the problems of developers and those who lent to them. Already there had been several bankruptcies among (mainly private) developers, and more were expected. The importance of cash flow was again asserting itself.

Cycles of boom and bust are by no means unknown in property. But at the beginning of the 1990s, longer-term issues were also emerging. For most of the post-war period, scarcity of space had helped property rents and values to grow, on balance, at a respectable rate. In the somewhat less restrictive planning environment of the late 1980s and the 1990s, it was open to question whether this scarcity factor would continue to apply to the same degree. If not, a rather slower pace of rental growth might be expected and property might come to be regarded rather less as a growth investment and rather more as an income investment and as secure backing for long-term debt.

That was how the picture looked in 1990. As we now know, the next three years were to shake the property industry to its foundations. Those questions we were asking in 1990 remain valid today. But the crash of 1990 to 1993 has posed a host of totally new questions, many of which are addressed in the later chapters.

One fact is certain. No longer is property regarded as an investment in isolation. Increasingly, returns from property need to be compared with those from other forms of investment such as government bonds and ordinary shares. "Property and Money" attempts to explain the financial and investment facets of property in a way that allows such comparisons to be made.

A few points of detail need brief explanation. The book falls into several parts. It starts with an explanation of the basic mechanisms of the property investment market and of property evaluation. This is followed by a section on the principles of property finance and on different types of property companies and their accounts. The accounting material has been expanded by a new chapter (Chapter 16) which introduces the important accounting changes of the last few years. Stock market launches and takeovers also fall within this section. Next comes a series of chapters on different types of short- and long-term finance for property. They are followed by a brief look at cycles in the property market and a review of changes in financing techniques in the post-war period up to 1990.

At this point appears a totally new chapter analysing the 1990-93 property market crash and its more important side effects. It is followed by a series of chapters which look in detail at some of these new problems for the property industry. The narrative portion ends, as in the first edition, with a brief look at the main players in the property market. At the end come a glossary and index. The index points to the sections of text where terms or concepts are explained, while additional terms are covered in the glossary.

The interest rates adopted in the examples are not necessarily intended to be those applying at the date of publication. Interest rates can change very rapidly and the rates chosen are generally selected for illustrative purposes only.

Tax rates also change rapidly — though not quite as rapidly as interest rates. As we are more concerned with the principles than the soon-to-be-outdated detail, we have stuck in our examples with the tax rates applying in 1990-91: a basic income tax rate of 25%, an upper rate of 40% and a corporation tax rate of 35% (in practice the corporation tax rate was 33% in 1996 and the basic income tax rate was shortly to reduce from 24% to 23%). As this edition was going to press, the July 1997 Budget radically changed the system of tax credits on dividends. The new regime, which supercedes the one described in the main body of the text, is outlined in a "stop press" entry in the glossary.

Again in the interests of simplicity, we have made no allowance for the cost of transactions in the examples, as this could sometimes obscure the principles. In practice, costs may be a significant factor. We have also tended to use the term "commercial property" fairly loosely to cover factories and warehouses ("industrial property") as well as shops and offices. It is thus used to cover the main types of property in which institutional investors are interested.

And lest a surfeit of facts and figures prove daunting, the text is enlived with Nick Newman's cartoons.

1
Characteristics of property

What gives property its characteristics as an investment? Right at the outset we have to recognise that investing in commercial property is a totally different business from investing in securities traded on a stock exchange. The returns from property can be compared with the returns from other forms of investment, but the mechanics of investment are different.

First, commercial properties are of large individual value. You can buy a few hundred pounds' worth of shares in ICI, but you are probably talking in terms of hundreds of thousands of pounds for a good quality commercial property and you may well be talking of many millions. So commercial property ownership tends to be concentrated in the hands of large investing institutions and companies, though some rich private individuals participate.

Second, property is not a standardised investment. One ordinary share in ICI is just like any other ordinary share in ICI, but no two properties can be identical in terms of location, structure, tenant and lease.

Third, property is not a pre-packaged investment. If you buy shares in ICI, the ICI management runs the company, produces the profits and pays the dividends. All you have to do is cash the dividend cheques. If you own a property, you have to manage it yourself or pay someone to do so on your behalf.

Fourth, property is an investment that can often be improved by active management. As a small shareholder, there is probably nothing you can do to improve the value of your share in ICI. But with a building you may be able to add value by buying in leases, refurbishing or even redeveloping the structure itself. This means it may be best suited to the class of investor who has a constant flow of new money to invest.

Fifth, property is an investment you can create for yourself by acquiring land or rights to land, erecting buildings on it and finding the tenants.

Sixth, it follows from the three previous points that some expertise is needed.

Seventh, there is no single market for commercial property. It may be traded at auction or sales and purchases may be negotiated by the major firms of estate agents and surveyors.

At any one time there will be a multitude of private transactions taking place at different locations. And it takes time to go through all the processes involved in a property purchase or sale — much longer than is the case with shares.

Eighth, information in the property market is often imperfect. The major estate agents know much of what is going on from the deals that pass through their hands. But it is not always easy to find out what is happening at a particular location, who owns the adjoining buildings, what nearby tenants are paying, and so on. The price at which transactions take place is not always made public — another way in which property differs from shares. Worse, it is not unknown for a certain amount of *disinformation* to be fed to the more gullible sections of the press. But the quantity and quality of information available is improving.

Ninth, while most commercial property is bought in the hope that it will produce an increasing stream of income and that its capital value will rise to match the rising rents,

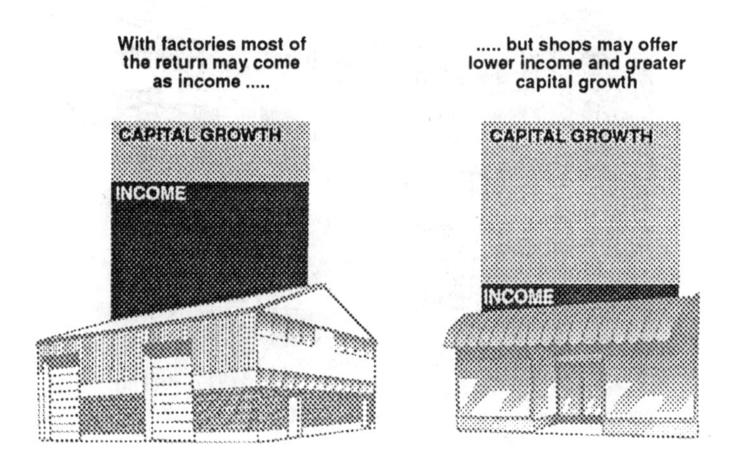

the income does not increase as a steady progression. Leases probably contain provision for the rent to be reviewed (upwards only) every five years. So even if rental values are rising rapidly, you have to wait five years from the start before the rent you receive can rise to reflect this fact. For this reason your return has to be adjusted to compare it with returns from dividends on shares, which may increase every half year.

Tenth, the physical structure of buildings may wear out and need replacing, but the land on which they stand does not. However, the value of a particular piece of land (and the building on it) can be affected by developments nearby, by planning consents or refusals and by pure fashion.

Eleventh, there are many different forms of title the property investor can have to his building and the land on which it stands. He may own the building as a freehold. He may have a long leasehold interest which allows him to enjoy benefit of ownership until the ground lease expires. He may have a short leasehold interest which constitutes a wasting asset; he enjoys the income meantime, but is left with nothing when the lease expires.

Twelfth, it follows from the previous point that different types of property investment will suit different people or institutions according to their tax status. The buyer of a short leasehold has, notionally at least, to set aside enough of his income each year to replace (amortise) his original outlay on the investment by the time the lease expires. Short leaseholds may therefore be most attractive to pension funds which do not pay income tax, since they can amortise the investment from income which has not borne tax.

Thirteenth, there is another sense in which different kinds of property suit different investors. Industrial buildings (factories and warehouses) can be bought to show a fairly high yield, though the growth rate may not be enormous. Good quality shops probably provide a lower income yield, but may grow more rapidly in capital value. The tax and cash flow position of the purchaser can affect his choice.

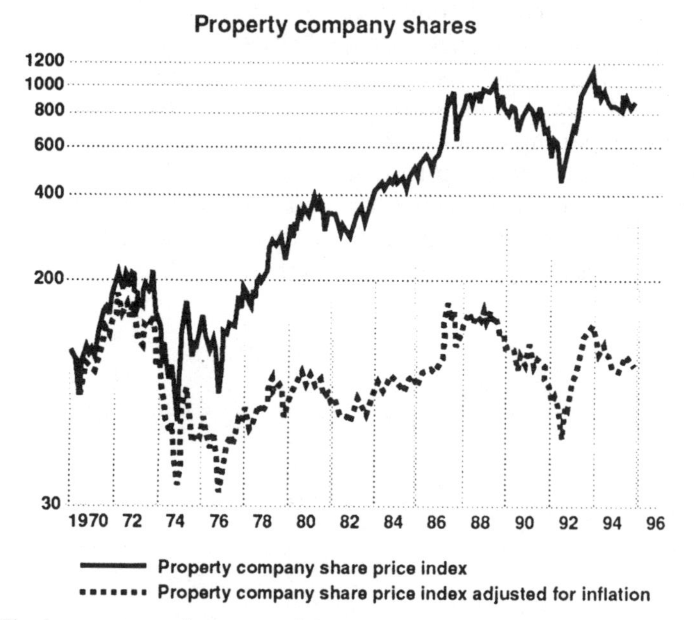

The long-term performance of property company shares, as measured by the Datastream index. The performance does not look quite so impressive when the values are adjusted for inflation (bottom line). Source: *Datastream*

Fourteenth, property, like almost all investments, carries risk in some form or other. The very best quality ("prime") properties will usually be in demand and in this sense the buyer faces little risk. But there is still the risk that he may find he paid too much, even for the best. Lower grade "secondary" properties may provide a considerably higher income and can sometimes demonstrate high capital growth. But there is probably more risk; if the tenant goes bust, the property might be difficult to re-let.

Fifteenth, rental income from good quality property is, however, relatively low risk. Companies must pay the rent long before they think about paying a dividend on their shares. And with a well sited property, if one tenant gets into trouble it should be possible in "normal" market conditions to find another without undue delay.

Finally, property is often presented as a long-term investment. In practice, owners often hold on to their properties for many years, but there is no reason why this should necessarily be so. However, values do not rise in a steady progression. There are strong cycles in the property market (see Chapters 44 and 45) which do not usually exactly match the cycles in other forms of investment. Growth in rents and values may be very slow for several years, then suddenly shoot up. The long-term investor experiences the upswing as well as the down-swing of the cycle.

2
What we mean by yield

Most forms of investment ultimately come back to the question of income. The investor who buys a plot of vacant land for £100,000 and sells it for £200,000 may not be thinking in terms of income. He has obtained his return in the form of a capital profit on the sale of the land. But somebody at some point will need to put that land to work to earn an income: whether he farms it to earn an annual profit or builds office blocks on it which can be let to produce a rent.

The capital value of land or property (which is just another way of saying "the price") is thus generally related to the income it produces or could produce (we will ignore housing for the moment, where the considerations are rather different). The buyer of a revenue-producing property investment — a tenanted office block, say — is effectively paying a capital sum today in return for the right to receive a stream of income in the future.

If the property is a freehold, he has the right to the income in perpetuity, though it will not necessarily remain the same. The investor hopes the income will increase, though at some point the building may become obsolete and the income will decrease or cease altogether as tenants choose more modern properties instead. At this point the owner will need to lay out further funds to refurbish or redevelop it so that he still has a building for which a tenant is prepared to pay rent.

The yield on a property is usually defined as the initial return — in the form of income — that an investor receives on the money he lays out to buy a property. To illustrate the principle we are talking for the moment only of buildings let at a current market rent. So, say the investor pays £1m for an office building let on a full repairing and insuring lease (where the tenant bears most of the running and maintenance costs) at a rent of £60,000 a year. He gets an initial return of 6% on his money. The sum to calculate the yield is simply:

$$\frac{\text{INCOME}}{\text{PRICE PAID}} \times 100$$

or in this case:

$$\frac{£60,000}{£1,000,000} \times 100 = 6\%$$

It is the same sum as is used to calculate the gross yield you receive on a share or a bond when you buy at the market price.

Why should an investor accept a return of only 6% on the property even at times when he might get a perfectly safe return of, say, 10% on a fixed-interest government stock: a gilt-edged security?

The answer, of course, is that he expects the rental income from the property to grow as the general level of rents rises, whereas the interest income from the gilt-edged security will remain exactly the same throughout its life. It is worth accepting a lower yield today in return for the expectation of a rising income in the future.

But the income is not the whole of the story. If the rents from the office block rise, not only does the owner get a higher income. The capital value of his investment (the price somebody would be prepared to pay for it in the market) should rise as well. Suppose that when the rent on the office is reviewed after five years it increases to £97,000 — a rate of growth of about 10% a year compound. If an investor would still be prepared to buy the office block at a price that showed a 6% return on his money, he could now pay £1.62m. This is the value to which the office block has risen.

The original owner thus has a potential capital profit of £620,000 as well as the income he has received over the five years: a total return of over 15% a year compound on his original £1m outlay. The reason 6% a year income and 10% a year capital growth do not add up to a 16% overall return is the five-year rent review pattern in property, which means a landlord has to wait five years for his increased income.

But remember that this 15% is to some extent a notional or "paper" return; the owner is only sure of it if he actually sells

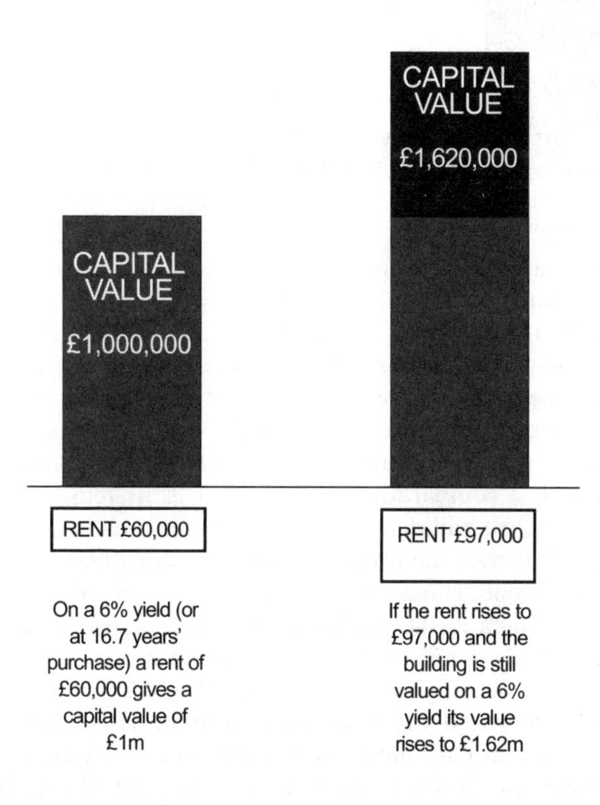

the building. If he holds on to the office block at the new higher rent, the income return alone on his original £1m investment has now risen to 9.7% a year.

Note that we tend to talk of "yield" when referring to the return the investor receives in the form of income, and "total return" or "overall return" for the return he gets from the combination of income and appreciation in the capital value of the property.

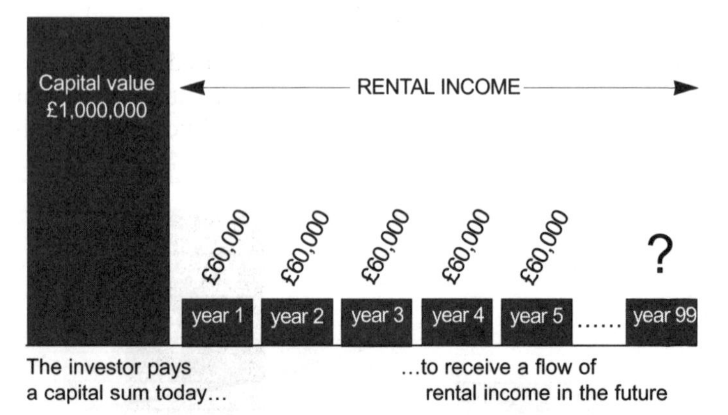

The value of a property is often expressed in terms of its yield. Remember, we are still talking for simplicity of buildings let at the current market rent. Valuation becomes more complicated in the case of a building currently let at a rent below the going market rate — a reversionary property. And it becomes a lot more complicated when a building is over-rented: let at a rent above the market rate (but since over-renting is a comparatively new and possibly temporary phenomenon, we will defer looking at it until the later chapters which deal with the property crash of 1990-1993).

Our freehold building producing a rent of £60,000 a year — which is the market rent — and valued at £1m has been valued on a 6% yield basis. Another way of saying this is that it has been valued at "16.7 years' purchase" of the rent. In other words, if you buy a building at a price showing you a 6% return on your outlay, you are paying 16.7 times the rent it initially produces. If you buy on a 5% yield you are buying at 20 years' purchase, and so on. The "years' purchase" is a similar concept to the "price-earnings ratio" yardstick used to express the rating of a share in the stock market. Both relate the price of an investment to the revenue it earns, though the price-earnings ratio is applied to earnings after tax. The sum to give you the years' purchase for a property is simply:

$$\frac{100}{\text{YIELD}}$$

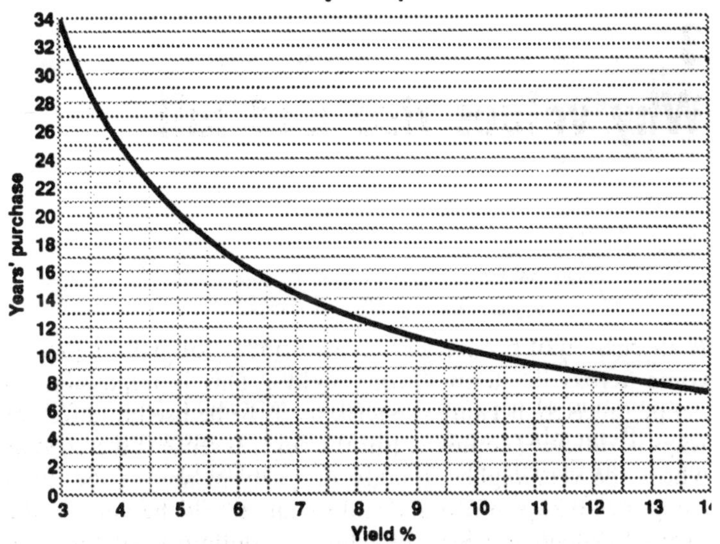

A graphical method of showing the relationship between property yields and years' purchase of the rents. Buy a property to show a 5% initial rental yield and you are buying at 20 years' purchase. And so on. In the age of calculators it is simpler to convert from yield to years' purchase by dividing the yield into 100.

or in this case:

$$\frac{100}{6} = 16.67 \text{ YEARS' PURCHASE OR "yp"}$$

The yield on which investors are prepared to buy properties let at the current market rent depends primarily on the rate at which rents (and therefore capital values) are expected to grow. The faster the growth you expect, the lower the yield you are prepared to accept at the outset. Other factors such as risk come into the equation, but in general it holds true. Thus top quality shops (whose rents are generally expected to increase rapidly) often command the lowest yields and industrial buildings (factories and warehouses which traditionally have a lower growth rate) command the highest.

3
Why values rise and fall

In discussion of property as an investment you will frequently come across the term "prime yield". This is impossible to define precisely because the term tends to be used in different ways in different contexts. But the general sense is the yield at which the very highest quality property in any of the main categories (shop, office and industrial) would be valued: the property which is expected to show the optimum combination of high growth, low risk, easy marketability and so on.

Thus at March 1990 estate agents Healey & Baker quoted the following prime yields (they are probably more illustrative of post-war patterns than post-crash figures would be):

Shops	5.0%
Offices	5.0%
Industrials	7.75%

These are the very lowest yields that investors are prepared to accept for the class of property. They may be useful as a yardstick. But since only a fairly small proportion of investment-quality property in any of the categories would qualify as prime, most buildings held as investments by the insurance companies and pension funds would be valued on higher yields. Average yields for investment-grade property do not always follow the same trend as prime yields, so the latter need to be used with some caution.

Now let us return to the £1m office block producing a rent of £60,000 a year. It is of good quality but not quite prime and is therefore valued on a 6% yield.

If something should happen to increase the rental growth prospects of this particular property, all else being equal we might expect the yield on which it is valued to drop. An investor would now be prepared to accept a yield of below 6%, because the growth in income looks like being faster than was earlier expected.

This might happen in two main ways. Perhaps there is a general shortage of office property in the country and the rents that landlords can ask are therefore rising more sharply. Or perhaps this particular building has become more attrac-

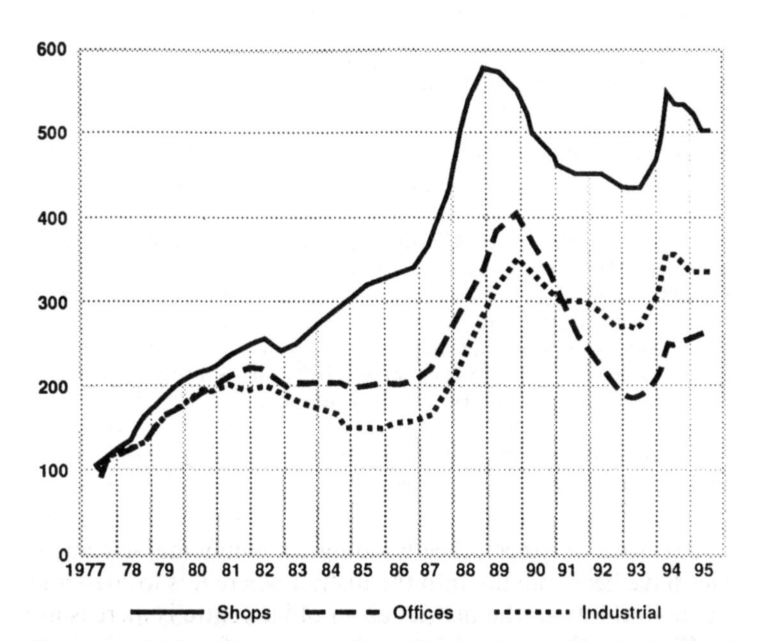

How capital values have moved since 1977 for the three main classes of business property. The greatest value growth has been provided by shops, a fact recognised by average yields that are normally lower than for offices or industrials. The performance of offices has been severely dented by the sharp fall in values since 1990. Industrial property has combined a reasonable capital performance with a generally higher income yield than the other classes. Source: *Hillier Parker; Datastream*

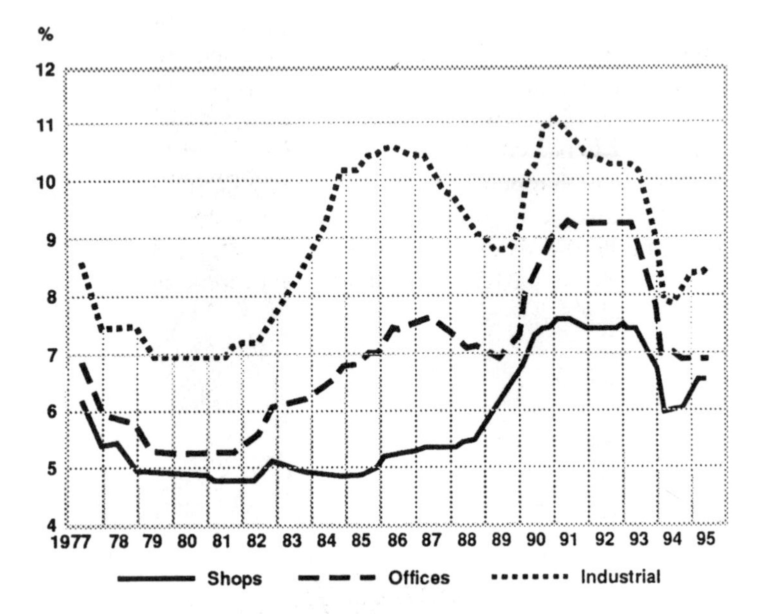

Of the three main classes of business property, shops normally change hands on the lowest yields and industrial property on the highest. Usually, offices change hands on higher yields than shops (though the position may be different for City of London offices at the height of a boom). The cycles for the different classes of property may also be different. In the late 1980s shop yields began moving up towards those for offices as investors worried about the ending of the 1980s consumer boom before they became concerned about the effects of recession and oversupply on office values. Source: *Hillier Parker; Datastream*

tive relative to other office buildings: possibly major companies have been moving into the district where it is located and the desirability of the district as an office centre is increasing.

Whatever the reason, the results are gratifying for the building's owner. If the yield drops from 6% to 5.5%, the number of years' purchase rises from 16.7 to 18.2. Instead of being worth £1m the building is now worth £1.09m (the £60,000 rent multiplied by the 18.2 years' purchase). The value has risen because of evidence of more rapidly rising rents.

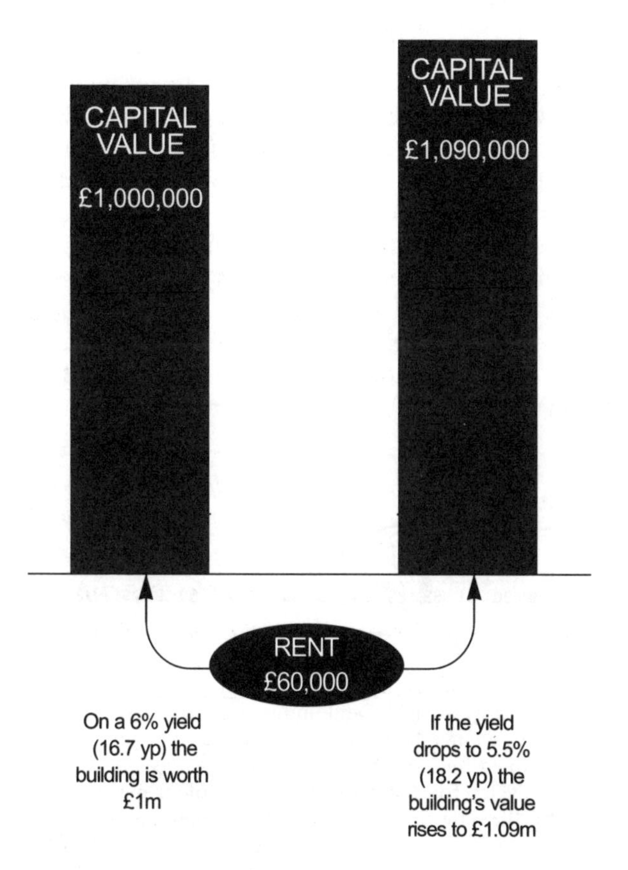

However, there is another way in which the yield might come down and the value therefore rise. Let us suppose that nothing has happened to change expectations of rental growth. But there is a growing shortage of completed and revenue-producing properties for the institutions to buy as investments. As a result there is too much money chasing too few investments, and the value of the investments will therefore tend to rise. This means that the yields come down.

The institutions may be piling into the market because returns on property look attractive compared with those obtainable on other forms of investment like government stocks. Perhaps yields on government stocks have come down. In theory, returns on different forms of investment should bear some relationship to each other and a movement in one is likely to trigger a movement in another. In practice, yields on property are often rather slow to respond to yields available elsewhere (see Chapter 44).

Or perhaps the institutions are simply following a current enthusiasm for property investment which has no very rational basis — they sometimes do. Whatever the reason, the yield on which our office block would be valued drops from 6% to 5.5%, though there is no justification for the change in the shape of faster rising rents.

For the property owner the immediate effect is much the same as if rents had been growing. The value of his property is £1.09m rather than £1m. But in the second case the increase in value is more suspect. The temporary enthusiasm among buyers could abate, the yield could rise to 6% again and the value would drop back to £1m. This phenomenon has been seen several times in the property market in the past. A rise in value based on income — rising rents — is far more secure.

There is one other phenomenon to look out for. Suppose our investor bought his office block, producing an income of £60,000 a year, for £1m. And suppose that rents, instead of increasing by about 10% a year compound as he had expected, rise by only 4% a year.

When the rent is reviewed after five years it will rise from £60,000 to £73,000. But at that point a valuer takes a rather different view of the property. Rental growth has been lower than initially expected and future rental growth prospects are not particularly good. The property is no longer worth buying on a 6% yield. A prospective purchaser might want an initial yield as high as 7.5% because of the more limited growth potential.

A 7.5% yield is equivalent to 13.3 years' purchase. Look at the effect on the building's owner. The rent has gone up from £60,000 to £73,000. But a buyer would now only pay 13.3 times that £73,000. This works out at £971,000. Despite the rise in rent, the value of the office block has actually dropped.

This is what can happen if the quality of a property as an investment is downgraded. The art of successful property investment is to pick properties which are likely to be upgraded.

For the moment the main points to grasp are therefore:

- The yield on a property by and large reflects the expectations of growth. As a general rule, the lower the yield the higher the growth expectations.
- When rents rise, all else being equal the capital value of the building rises.
- When yields fall, all else being equal the capital value of a building rises. When yields rise, all else being equal the capital value falls.

In a greatly simplified form, these are are the basic mechanisms of the property investment market.

4
Valuing future rental increases

So far we have looked at the basic mechanism of yields in the property investment market: how capital values of tenanted buildings rise and fall with changes in the yield basis of valuation.

But the example we took was of a building recently let at a current market rent: the rent that the building produces is the same as the rent which it would command if offered, vacant, on the market today (the current rental value).

With most commercial buildings, the lease allows for the rent to be reviewed only at five-year intervals. This means that if a landlord lets a building and rents subsequently rise, he has to wait until the rent review for his higher income.

But this does not mean that the capital value of the building also remains static until the rent review. All else being equal, the capital value will rise to reflect the higher rent expected in the future. This sometimes causes confusion when talking of the yield on which the building is valued.

Let us go back to our earlier example of a building initially let at £60,000 a year and, on a 6% yield basis, valued at £1m. Rents then rise very sharply and after two years the rental value has risen to £90,000 — this is the rent that it would fetch if it were vacant and being re-let or let for the first time.

Because the first rent review does not take place until five years after the original letting, at this point the landlord still has to wait another three years before he can negotiate the higher rent with the tenant. Of course, if rents continue to rise, the rent negotiated at the end of the five years will be higher than £90,000. But for the moment — two years into the lease

— the current rental value of the building will be the basis for valuation.

The valuer then looks at the income the building produces in two ways. There is the rent it currently produces, which has a value. And there is the additional rent it will produce in three years' time, and this expectation also has a value.

In practice there are different approaches, though they will give much the same result. We are concerned mainly with the principle, so we will take the simplest method.

First, the building currently produces £60,000 a year, and on a 6% yield basis (16.7 years' purchase) the right to receive this income in perpetuity is worth £1m.

But in three years' time there will be an additional £30,000 a year coming in. Again on a 6% yield basis, the right to receive £30,000 a year in perpetuity would be worth £500,000.

But £1 today is worth more than £1 for which you must wait three years. The reason is that if you had the £1 today you could put it to work to earn interest, so that it would have accumulated to more than £1 after three years. So the extra £30,000 a year the landlord will start picking up from year five is worth less than £30,000 in today's money. Or looked at a different way, the additional capital value stemming from the higher rent in the future is not worth £500,000 in today's money. It has to be "'discounted" back to reach a present value.

The rate of interest a valuer will use to discount this additional value is by convention the same as the yield on which the current income from the building is valued — 6% in our example. If we discount the £500,000 extra value at 6% for the three years till the rent review, we get a present value of £420,000. In other words, at 6% compound interest £420,000 would grow to £500,000 in three years. Therefore, £500,000 in three years' time is worth £420,000 today at a 6% discount rate.

So the value of our office block, two years into the lease and when rental values have risen by 50%, is the £1m capital value of the income which it currently produces plus the £420,000 discounted value of the additional income expected in the future: a total of £1.42m.

If we then express the current rent — £60,000 — as a percentage of the capital value of £1.42m, it comes out at 4.23%:

$$\frac{60,000}{1,420,000} \text{ X } 100 = 4.23\%$$

This is the initial return (also called initial yield) that a buyer would get on the building if he paid £1.42m for it today. But it does not by itself tell us anything about the market level of yields for similar buildings let at a current market rent, which is the yield basis the valuer will have used in his valuation of the building. As the sums show, the yield basis used throughout has been 6%.

At 6% a year compound interest, £1 will grow to £1.191 after 3 years

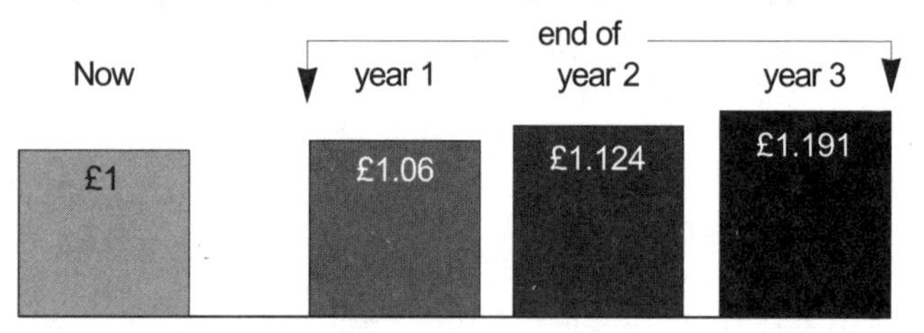

So at 6% interest, £1 received in 3 years' time is worth only £0.8396 today

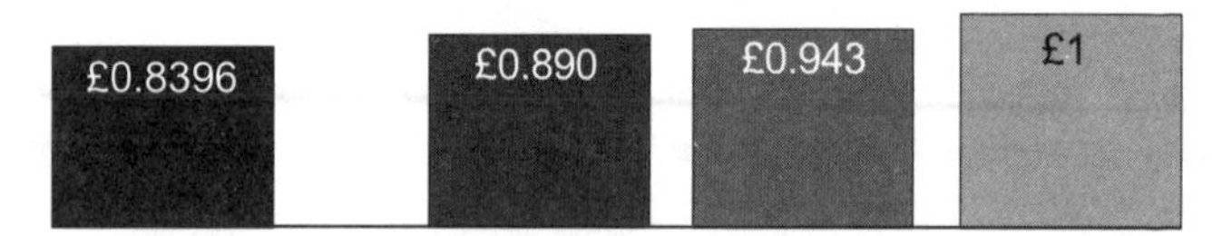

In other words, £0.8396 is the sum that will grow to £1 in 3 years at 6% compound interest

So, when talking of reversionary buildings of this kind, commentators often refer to the "equivalent yield". This is the underlying yield basis — in technical terms, the "all-risks yield" — used in the calculations rather than the immediate yield that a buyer would get on his money if he bought a reversionary property at today's price. The equivalent yield tells us something about the way in which the valuer rates the quality and growth prospects of the investment — it is a building he would value at 6% if let at a current market rent. The actual yield to the purchaser does not tell us very much at all about the building; it simply tells us what immediate income the buyer will receive.

The reversionary element in the valuation has one or two side effects. It means that, if rents are rising very rapidly, the capital value of the tenanted investment does not rise quite so fast because the increase has to be discounted back from the point of the next rent review. It also means that the value of the building would — all else being equal — continue to rise modestly each year up to the rent review even if rental values did not increase any further in years three, four and five of the lease.

This is because, as each year passes, the additional rent expected at the fifth year is being discounted for a shorter period. At a 6% discount rate, £500,000 three years off is worth £420,000. But £500,000 two years off is worth

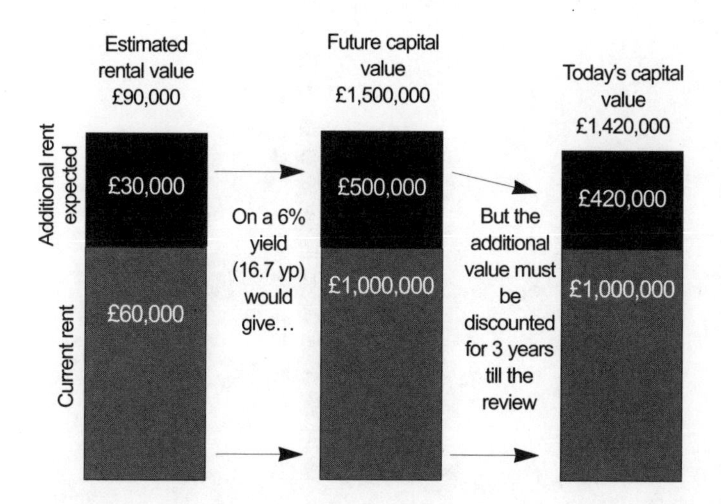

The building, valued on an equivalent yield of 6%, is let at £60,000 but would fetch £90,000 today if vacant. The rent review is three years off, so the capital value of the expected additional income must be discounted for three years at 6% to arrive at today's value.

£445,000. So even if rents did not change further between years two and three, and provided the 6% yield basis remained constant, the value of the building would rise from £1.42m to £1.445m between these years: an increase of about 1.75%.

It is also worth noting that a building does not have to be tenanted and rent-producing to be valued on the basis of its rental value. If it is capable of being let, the valuer probably works on the basis of his estimate of rental value, making some deduction from the final figure for the current lack of a tenant. Thus a building may be rising in value while it is untenanted if rental levels in the area are rising.

5
Evaluating the asking price

The valuer of a property normally tries to calculate what a property is likely to be worth by reference to recent transactions in comparable properties. But there are other approaches to what a property is worth and the return it is likely to provide. For example, what is it worth to a specific buyer and what return might a specific buyer derive from it, given the assumptions he chooses to make? This is the investment analysis approach.

This is where discounted cash flow (DCF) sums come into the picture. They are particularly useful in allowing a prospective purchaser to calculate what price he could afford to pay for a property to show him the overall return he requires. And they allow a property owner to calculate the actual return he has derived from a property in the past or estimate the return it will show him in the future, based on his own projections of income, capital values and possible future expenditure on the property.

How does this differ from the valuer's approach? After all, the value of a property is generally a reflection of the income streams it is expected to produce in the future. But the traditional valuer is not having to specify all of his assumptions. His expectations of growth, risk and so on are all implied in the yield — the "all-risks yield" — on which he chooses to base his valuation. And this yield in turn reflects yields on comparable buildings in the property market.

Anyone undertaking a DCF calculation has to be far more precise as to his assumptions. He has to specify the flows of funds from and into the property. And the rates of interest he uses in his calculations reflect market rates of interest rather than yield patterns specific to the property market.

How does this work in practice? First, remember the basic principle that £1 in the pocket today is worth more than £1 receivable in the future because, if you had the £1 today, you could be putting it to work to earn interest. And remember that it follows from this that £1 receivable in the future has to be "discounted" to find what it is worth today. The crucial factor is the rate of interest at which you choose to discount. If you thought you could currently put £1 to work to earn 10% interest, you might discount £1 receivable in the future at a 10% rate of interest.

At a 10% rate of interest, £1 receivable in three years is worth about £0.7513 today. This is the same thing as saying that £0.7513 is the sum which will grow to £1 in three years, at 10% a year compound interest. You do not have to work these figures out for yourself — there are plenty of compound interest tables available, or programmes for computers or pocket calculators.

So £0.7513 is the "present value" of £1 receivable in three years, at a 10% rate of interest. In the same way, you can

calculate the present value of a whole series of receipts and expenditures involved in a property investment, by discounting each for the appropriate period. The aggregate of these discounted values will be the "total present value" of the property, at the interest rate you have chosen and on the assumptions you have made. The total present value may then be compared with the price of the property. If it is greater, the property is a "buy". If lower, the property is overvalued from your viewpoint. An example makes the process clearer.

Suppose the asking price for an office building just let at a rent of £60,000 a year is £1m. In other words it is valued on a 6% yield. Your criterion for this sort of investment is that it must show you a total return of at least 14% a year. Should you buy it at the asking price?

Your own estimate is that the rental value will have risen to £90,000 a year by the first rent review in five years' time. You also reckon that in five years it will be valued on much the same yield as at present. So in five years its capital value will be £1.5m.

The next stage is to work out your receipts from the building, including the actual or hypothetical sales proceeds after five years. For simplicity we are ignoring purchase and sales costs — though they would in practice be allowed for in the calculation — and we are also following the common convention that rents are paid annually in arrear. In practice they are paid quarterly in advance.

The picture looks like this:

Receipts	£
End of year 1	60,000 rent
End of year 2	60,000 rent
End of year 3	60,000 rent
End of year 4	60,000 rent
End of year 5	60,000 rent
End of year 5	1,500,000 sales proceeds

Now each of the receipts from the property has to be discounted back for the appropriate period to reach a present value. At what rate of interest do you discount? You take your 14% target rate of return. At a 14% discount rate, a pound receivable in a year's time is worth £0.877, in two years' time £0.769, and so on. So £60,000 receivable in a year is worth £60,000 X 0.877, or £52,631.

	Receipts £	Discount Factor (at 14%)	Present Value £
Year 1	60,000	0.877	52,631
Year 2	60,000	0.769	46,168
Year 3	60,000	0.675	40,498
Year 4	60,000	0.592	35,525
Year 5	60,000	0.519	31,162
Year 5	1,500,000	0.519	779,053
TOTAL			**985,037**
Asking price			1,000,000
DIFFERENCE			**–14,963**

In this case the present value of all the receipts from the investment, including actual or hypothetical sales proceeds of £1.5m at the end of the day, is £985,037 at your chosen target rate of return of 14%. This means that if you bought the building at £985,037 and your assumptions on growth and end-value turned out to be correct, you would receive a return — an internal rate of return or IRR — of 14% on your outlay. So £985,037 is the most you could afford to pay.

But the asking price of £1m is above the present value you have arrived at by the calculation, and therefore the "net present value" — present value less the asking price — of the building to you on your chosen assumptions is negative to the tune of £14,963. If you paid the asking price the return would be below your 14% target. Either you don't buy, or you try to negotiate the price down to the £985,037 you can afford. See illustration on page 37.

Of course, a sum of this kind may well throw up a positive net present value — the discounted value of the receipts is above the asking price — in which case the building is cheap according to the criteria you have chosen.

There are a number of virtues to the discounted cash flow approach, which can be used in many different forms. In essence there are three elements to a calculation:

- The price asked for a property (plus any additional expenditure expected)
- The receipts expected
- The rate of return

Given any two of these, it will be possible to calculate the third. In our example we started with the rate of return required and the assumptions as to income and future capital value, and from this were able to work out the price that could be paid for the property.

But equally we could have started with the asking price for the property and the expected rents and future capital value, and from this could have calculated the rate of return.

Or we could have taken the asking price and the required rate of return, and calculated what growth in rents and capital values the property would need to achieve to justify a purchase. The calculation might well show that the growth assumptions implicit in the asking price were unrealistic. We look in the next chapter at some of these alternative approaches.

6
Calculating the actual return

We have looked at how discounted cash flow (DCF) sums can be used to calculate the present value of a property to a purchaser under given assumptions and projections: the net present value calculation. This time we are going to do the sum the other way round to calculate the internal rate of return (IRR) from a property.

The internal rate of return is probably the best measure of the total return an investor gets on his money. The calculation does make certain assumptions — if receipts from the investment are reinvested, they can only be reinvested at the IRR itself — but we do not need to bother too much with the technicalities to grasp the principle.

First, why is an IRR calculation necessary? At the simplest level it is needed because you cannot very easily calculate the return from a property on a rule-of-thumb basis. If you have a property yielding 6%, whose rental value and capital value grow by an average annual compound rate of 10% over the next five years, there might be a temptation to assume your total annual return would be 16%: the 6% of income plus the 10% of capital growth.

In reality the return would not be as high as this, because you must allow for the fact that none of the increase in rental value is reflected in the rent payments you receive until the rent is reviewed at the end of the fifth year. Your actual overall return — adopting the usual convention that rents are paid annually in arrear — would be just over 15%. Remember that £1 in the future is worth less than £1 today.

Then, too, the picture may be complicated by additional expenditure on the building during the period you are measuring. And, indeed, you may want to calculate the effect on your returns of spending additional money on improving your investment. Take the following set of circumstances.

An investor owns an office block currently valued at £1m and let at an up-to-date market rent of £60,000 a year (a 6% yield). By the time of the review at the end of year five he expects the rental value to have risen to £90,000.

But suppose he reckons that at the end of year three he will have the funds to improve the building by spending £150,000 on, say, installing an up-to-the-minute lift. If he does this he

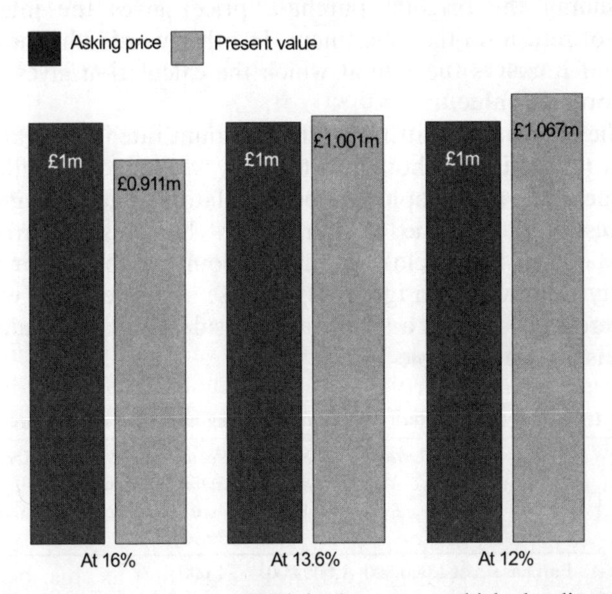

The internal rate of return (IRR) is the rate at which the discounted present value of the property (on your chosen assumptions) equals the price you pay for it. We are assuming a property with an asking price of £1m, producing £60,000 a year which is likely to rise to £90,000 in five years. If you want a 16% return, the property is too expensive. If you are only looking for 12%, it is cheap. If you pay the asking price of £1m, it will show you an IRR of about 13.6% on the assumptions you have made. The IRR is arrived at by trial and error (but a computer programme helps a great deal in the calculation).

expects that the rental value of the improved building at the review at the end of year five will be £100,000 rather than £90,000. He also hopes that the uprated building might in future be valued on a 5.5% yield (about 18.2 years' purchase) rather than the original 6% (16.7 years' purchase).

Will he improve his rate of return by spending the extra money?

First we need to calculate the rate of return on the building as it stands, given the owner's assumptions. This involves discounting back all outgoings and receipts for the appropriate periods, as explained in the net present value calculation. The discount rate at which the discounted value of all receipts equals the discounted value of all outgoings (including the original purchase price) gives the internal rate of return on the investment. In other words, the internal rate of return is the rate at which the calculation gives a nil net present value.

The only way of hitting on this discount rate is by trial and error ("iteration"), though fortunately we can get a computer or one of the more sophisticated calculators to go through the process for us. The answer, as it happens, is around 13.6% — or this is close enough for our purposes. For simplicity we have again ignored purchase and sale costs, which in practice would be a significant consideration. The calculation is shown in Table 1.

Table 1: Internal rate of return calculation assuming no further expenditure

		Outlays £	*Present value* £	*Discount factor (at 13.6%)*	*Present value* £	*Receipts* £
At start	Purchase cost	1,000,000	1,000,000	1.000		
At end of						
Year 1	Rent			0.880	52,817	60,000
Year 2	Rent			0.775	46,494	60,000
Year 3	Rent			0.682	40,928	60,000
Year 4	Rent			0.600	36,028	60,000
Year 5	Rent			0.529	31,715	60,000
Year 5	Sale value			0.529	792,866	1,500,000
TOTALS			**1,000,000**		**1,000,846**	

Now we can do the same sum on the assumption of £150,000 expenditure at the end of year three, a rent of £100,000 after the review and an end valuation of £1.82m based on a 5.5% yield or about 18.2 years' purchase. The discount rate at which discounted outlays and receipts now roughly equal each other is, we discover by trial and error, 15.17% (see Table 2).

Table 2: Internal rate of return calculation assuming improvements to building

		Outlays £	*Present value* £	*Discount factor (at 15.17%)*	*Present value* £	*Receipts* £
At start	Purchase cost	1,000,000	1,000,000	1.000		
At end of						
Year 1	Rent			0.868	52,097	60,000
Year 2	Rent			0.754	45,235	60,000
Year 3	Rent			0.655	39,277	60,000
Year 3	Improvements	150,000	98,191	0.655		
Year 4	Rent			0.568	34,103	60,000
Year 5	Rent			0.494	29,611	60,000
Year 5	Sale value			0.494	898,203	1,820,000
TOTALS			**1,098,191**		**1,098,526**	

So what this sum throws up is the fact that the expenditure of £150,000 at year three will be well worthwhile if the property owner's assumptions turn out to be correct. His return on the total capital invested in the property stands to rise from about 13.6% to almost 15.2% over the five-year period.

This is the sort of sum than a reasonably sophisticated property owner is undertaking when he attempts to decide whether a property will repay redevelopment or additional expenditure. Of course, in practice he will try to pin down the risk/reward balance by working the sums on a number of differing assumptions as to rental growth and sales value.

It is also the method that a fund manager will use in trying to quantify the past performance of his existing property investments. Since it takes into account the timing of expenditures it can give a very much more precise result than any rule of thumb method. The only problem is that no calculation can be better than the assumptions behind it and, as the examples show, the assumption as to the value of the building at the end of the period is a major factor. Even when analysing past performance, there is an element of subjectivity unless the building has actually been sold.

Finally, the internal rate of return calculation allows returns on different types of investment to be reduced to a common basis, so that genuine comparisons can be made. Take a share in an industrial company yielding 5% gross (before tax), where income and capital value rise at 10% a year compound over five years. Take a property yielding 5% which shows the same rate of growth in rental value and capital value. The share will show the higher internal rate of return because the increases in dividend income are received each year (or even each half year) whereas the property owner has to wait five years for the review till he gets any increase in income at all.

Most returns on investment are expressed on the assumption that the income is received once a year at the end of the year ("annually in arrear"). Thus, if an investment of £100 pays 10% a year interest, it is usually assumed this means it pays £10 interest at the end of each year.

In practice, the effective rate of interest may be different if the interest is paid at more frequent intervals or if it is paid in advance. Take a British government stock: a gilt-edged security bearing a fixed "coupon" or interest rate of 10%. In practice, the interest is paid twice yearly in arrear, so that an owner of a nominal £100 worth of the stock will receive £5 after six months and another £5 at the end of the year. He can thus invest the first £5 to earn interest for half a year, so he will earn more than £10 in interest over the year as a whole. Taking this factor into account, his effective rate of interest for the full year is 10.25%. Discounted cash flow analysis of the actual receipts allows the effective rates of return to be calculated.

By convention, property returns are also normally expressed on the assumption that rents are received annually in arrear whereas in practice they are received quarterly in advance. We have stuck with the convention in the examples in this book. However, effective yields on property are somewhat higher than the yields normally quoted if the actual pattern of rent receipts is taken into account. Modern valuation tables probably include years' purchase tables based on rents receivable quarterly in advance as an alternative to the more common "annually in arrear" figures.

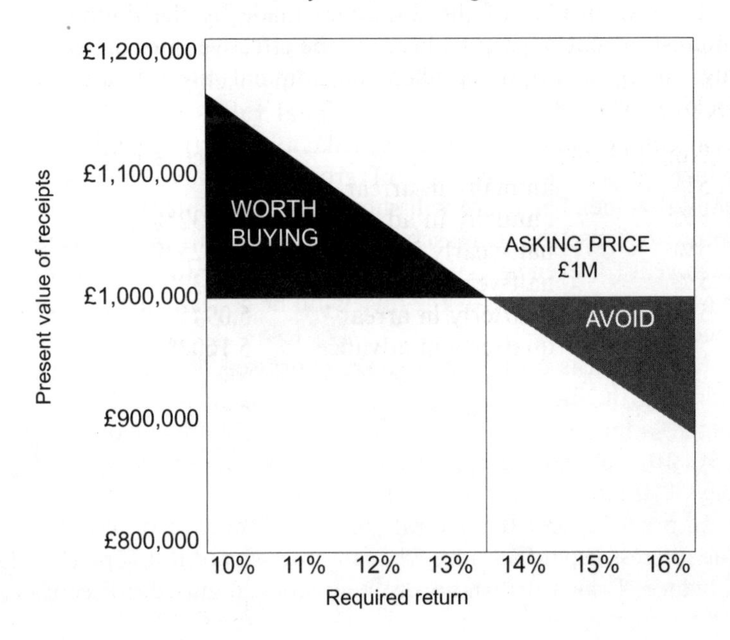

A property just let at a current market rent of £60,000 a year is on offer at £1m. You expect that after the review at year five the rent will rise to £90,000 and the value to £1.5m. You require a return of 14%. Should you buy? The answer is 'No". If you want more than 13.6%, on your assumptions the present value of the property is lower than the asking price. Its net present value is negative. However, if your required return is 13.6%, or less, the property is worth £1m or more to you.

In our comparison of a property and a share in an industrial company, both yielding 5% and showing 10% a year growth, the share shows the higher internal rate of return because the increased income from the property is not received until the sixth year and therefore has to be discounted for five years. In this example, the effective overall return on the share is about 0.6 percentage points higher than that on the property if we assume that the rents on the property are received quarterly in advance. The difference would be greater if income from the property were received annually in arrear.

To give an idea of the difference made by the timing of interest or rent payments, here are the effective rates on varying timing assumptions when the nominal interest rate (or yield) is 5%:

Nominal rate		Effective rate
5%	annually in arrear	5%
5%	annually in advance	5.2632%
5%	half-yearly in arrear	5.0625%
5%	half-yearly in advance	5.1939%
5%	quarterly in arrear	5.0945%
5%	quarterly in advance	5.1602%

7
Valuing leaseholds

The examples we have taken so far have all related to freehold properties. Valuation of leasehold properties is rather more complex, but not impossibly so if you take first principles as your starting point.

A freehold property, for most purposes, belongs "outright" to the owner. This does not mean that he can do what he likes with it in all circumstances. His freedom of action may be affected by the leases granted to the occupiers and by planning and other constraints, and in some circumstances the property might be compulsorily purchased from him against his will. But usually his ownership continues as long as he wishes.

A leasehold is a different proposition. The leaseholder enjoys most of the benefits of ownership of the building, but only for a specific time span. Technically, he owns a lease on the building and at some point the lease will expire. He also pays a rent under the lease.

Leaseholds can arise in a number of different ways. They can be created by a sale and leaseback transaction, used as a method of development financing (see Chapter 42). They may be created when a freehold landowner grants a developer a lease on his land for, say, 99 years, allowing the developer to put up a building at the developer's expense (this would be a "building lease"). The developer can let the resultant building to a tenant, but at the end of the 99 years the use of the land reverts to the landowner. At the same time the landowner acquires whatever building stands on it.

The leaseholder may pay a capital sum at the outset for his lease (a "premium"), though there will also be an annual rent which may be nominal or may be a significant amount. Or he may simply agree to pay an annual rent to the landowner, and this rent may be fixed for the life of the lease or may rise according to some formula.

The rent paid by the leaseholder to the freeholder is known as the "ground rent". Sometimes, as with leaseback development finance, it may represent quite a high proportion of the "rack rents" paid by the ultimate occupiers. When it is purely a nominal amount it is often known as a "peppercorn" rent. A fixed ground rent — the right to receive a set amount per year for, say, 99 years — is regarded as a fixed interest investment, and its value to an investor will change to reflect general movements in interest rates. But it should be very safe.

At the other end of the spectrum, the ground rent received under a leaseback financing deal is an equity type investment — it goes up with increases in the rental value of the property. Between these extremes of fixed interest and full equity participation, terms of leases and the ground rents payable can be structured to provide virtually any investment characteristics required. The freehold/leasehold structure is thus a popular way of providing profit participation and dividing the benefits of property ownership among a range of different parties.

While we have talked so far of a simple structure of freeholder/leaseholder, in practice the chain of "ownership" could be far more complex. The freeholder of a piece of land might be, say, a local authority which grants a long lease to an insurance company at a fixed ground rent. The insurance company in turn provides the finance for a building to be put up on the site by a developer, granting the developer a long lease at a ground rent which rises at five-year intervals in line with the rents the building brings in. The developer in turn grants an occupational lease at a market rent to a large industrial company which uses the building as its headquarters. And the industrial company itself might (if the terms of its lease allow it) subsequently let to a further tenant a proportion of the space which became surplus to its requirements — perhaps at a higher rent than it is paying to the developer.

Each of the parties — local authority, insurance company, developer, industrial company and subtenant — has an interest in the property. Each of these interests has different characteristics and can be separately valued. The one obvious constraint on structures of this kind is that nobody on this ownership ladder can grant a lease longer than the one he himself holds from the next body up the structure.

However, for purposes of illustration we will take a simple structure where a leaseholder pays a ground rent direct to the freeholder and in turn receives rack rents from the occupational tenant. But if the next person up the chain from the leasehold investor were not the freeholder, this person would be referred to as the "superior leaseholder" (or "head leaseholder" if he were the person holding the lease direct from the freeholder).

The first point to note about a leasehold interest is that it is a "wasting asset". It has a finite life which ends when the lease expires and the property "reverts" to the freeholder. At this point the leaseholder ceases to own any interest in the property. So while an investor who owns a freehold building will both receive a rent and have a building to sell when he wants to, the leaseholder gets his rental income from the building for a finite period, but has nothing left to sell once the lease expires.

It follows from this that (if rents remain static and yields are unchanged) a leasehold interest progressively loses its value during its life. The investor therefore has to allow for this fact in working out his returns and the price he can afford to pay for the investment.

In essence, this means that, from the rental income he receives, he must notionally or actually set aside an amount each year which will compound up to equal the price he originally paid for the investment by the time the lease expires. In other words, over the life of the investment he sets aside enough of the income to replace its cost by the time it becomes worthless. The return he gets from the leasehold building must therefore be worked out on the income he receives after making this provision. But note that the owner of a freehold can sell it at any time and afford to buy a similar property out of the proceeds. The owner of a leasehold, when the lease expires, will have set aside enough to replace its purchase cost but (assuming rents and values have risen) not enough to buy another identical investment with 50 years to run.

At this point we need to return to the principles of property investment. When an investor buys a freehold property he is effectively paying a capital sum for the right to receive a future flow of income. In the case of a freehold he has the right to receive the income for ever ("in perpetuity") though it may, of course, rise or fall over the years. With a leasehold investment the "owner" (the leaseholder) has a right to receive an income for a finite period.

So imagine two properties, identical except for the fact that one is a freehold and the other a leasehold with 50 years to run. For simplicity, assume that the ground rent paid by the leaseholder to his freeholder is a peppercorn — a nominal amount which can be ignored for the purposes of valuation.

Assume, also, that each property brings in £1m a year of rent, and that the freehold is valued on a 5% yield or at 20 years' purchase of the rent, making it worth £20m. How do the values of the two buildings compare?

One answer is that you apply the same initial 5% yield to the leasehold (there are varying views on this point, and a somewhat higher yield might be more common), but adjust for the fact that part of the £1m a year of rent you receive must be set aside to replace the price you paid for the investment by the time the lease runs out in 50 years.

What we are describing here is the traditional valuation method. Of course, you can perform a discounted cash flow calculation, working out the present value to you of the rental flows you expect to receive from the leasehold at your chosen rate of return and this is the way many property investors would approach the problem, particularly with very short leaseholds. But the traditional method demonstrates many of the considerations involved.

So, according to traditional valuation techniques, what sum do you need to set aside each year to build up after 50 years to a capital sum that equals the purchase price? We will cheat a bit at this point and assume that the purchase price for the leasehold was £17.68m (how the figure was arrived at will become clear later).

The normal way of providing a "sinking fund" (a fund which builds up to equal the original outlay by the time the lease expires) is to buy one from an insurance company. In return for an agreed annual payment, the insurance company contracts to provide the requisite sum at the end of the day. But the rate of interest it assumes in its calculation is likely to be low — it is offering a risk-free contract. In practice, of course, the buyer does not need to take out a sinking fund, but this sinking fund calculation demonstrates the valuation approach.

Suppose the rate of interest assumed in the sinking fund is 4%. From compound interest tables — using the "annual sinking fund" table — you can find out that, at an interest rate of 4%, annual payments of £0.0065502 compound up to £1 in 50 years. So to provide £17.68m in 50 years you need to set aside each year £17.68m multiplied by £0.0065502. It comes out at £115,830. This will be the annual cost of contributing to the sinking fund.

Thus, out of the £1m annual rent from the leasehold building, you need to set aside £115,830 simply to replace your capital by the time the lease runs out. This leaves you with only £884,170 a year from the rents (we are assuming for the moment that you are a non-tax payer, like a pension fund). Apply a 5% yield or 20 years' purchase to a rent of only £884,170 and you get a capital value of £17.68m (hence the figure we picked earlier for the value of the leasehold).

$$£1,000,000 - £115,830) \times 20 = £17.68\text{m value}$$

The equivalent freehold, remember, could be worth £20m.

In fact, the sum will normally be done in a slightly different way. Valuation tables include "dual rate tables" for leaseholds, which allow you to pick both the yield you are using in the valuation and the (different) rate of interest at which the sinking fund builds up. In our example we are assuming a 5% yield and a sinking fund building up at 4% interest. The relevant table shows that, for a 50-year leasehold, the rate of years' purchase is 17.6834. Multiplied by the £1m rent, this gives a capital value of £17.68m, which is the same answer as the earlier method gave.

So far so good, but unfortunately there is another serious complication. We have talked so far in terms of an investor who is a non-taxpayer. For a taxpayer, however, the sums work out differently. This is because he gets no tax relief on the eroding value of his lease.

Thus, he has to set aside amounts for the sinking fund out of income that has borne tax. If his tax rate is 35% (as tax rates change fairly frequently, we are sticking for simplicity with the rates applying in 1990-91) he will need £1.538 of rental income for every £1 he has to put into the sinking fund (£1.538 less 35% tax gives £1 net). So a £115,830 sinking fund contribution will actually absorb £178,200 of the gross rental income from the property. Therefore, when we take tax into account, we come up with a lower value for the property.

Fortunately, there are dual rate tables which show the effect of different rates of tax. Again assuming a 5% yield and a sinking fund compounding up at 4%, at a 35% tax rate the tables show that the years' purchase to apply for a 50 year lease would be 16.6452. On these assumptions our leasehold building producing £1m a year of rent would thus be worth £16.65m.

There are a number of consequences of this tax effect. A leasehold investment will tend to be worth more to a non-tax payer or a low-rate tax payer than to someone paying a full tax charge. But the differential is not the same for all lengths of lease.

With a medium-term lease (50 years or more, as in the example) the difference in value to different classes of investor is comparatively modest. But very short leaseholds (with five years or so of life, say) are a lot more valuable to the non-taxpayer than to the tax payer. This is because the contributions to the sinking fund will absorb a far greater proportion of the income from the property, since the sinking fund only has five years in which to build up to a sum that replaces the original capital. Clearly, there is a great advantage in these circumstances in being able to make the sinking fund payments out of untaxed rather than taxed income.

If we take the original example of a leasehold investment producing £1m a year, but with only five years of the lease to go, on a 5% yield and with a sinking fund at 4%, the tables would give a value of £4.17m (4.1732 years' purchase) for a non-tax payer. For the same investment, the tables would show a value of only £2.99m (2.9936 years' purchase) at a 35% tax rate. In practice, very short leaseholds will probably be evaluated with discounted cash flow techniques, with the buyer looking for a return related to the gilt yield.

Not surprisingly, non-taxpaying investors tend to make the market in very short leaseholds, whereas longer leaseholds are open to a wider range of investors.

For simplicity we have taken a leasehold with a peppercorn ground rent, but the same valuation principles apply in cases where a significant ground rent is payable. Suppose the rack rents from our leasehold investment were £1.2m and the leaseholder paid a ground rent of £200,000 to the freeholder. The leaseholder enjoys a net rent of £1m, and this is the rent on which the leasehold investment would be valued. It would thus be worth exactly the same as the otherwise identical leasehold receiving rents of £1m and paying only a peppercorn ground rent.

However, leaseholds resulting from sale and leaseback development financings cannot satisfactorily be valued on the same basis. Suppose the leaseholder receives rents from the occupational tenants of £1m, but pays a ground rent of £700,000 to the institutional freeholder. Suppose also that the ground rent to the institution rises at each rent review to 70% of the rack rental value.

The leaseholder's net rental income is currently £300,000. But it is a relatively risky "top slice" income. If a tenant went bust and the building became half vacant, the leaseholder would generally still have to pay the £700,000 ground rent to the institutional freeholder even though he was receiving rack rents of only £500,000 from the half of the building that was still tenanted. Also, leaseholds producing top slice profit rents of this kind are not always very readily saleable — they are likely to be worth more to the freeholder than to anyone else, and the freeholder may thus be virtually the

Income Tax 35% | Years' Purchase | Sinking Fund 4%

	RATE PER CENT							
YEARS	4	4.5	5	5.5	6	6.25	6.5	6.75
51	20.1495	18.3053	16.7704	15.4729	14.3618	13.8640	13.3996	12.9653
52	20.3239	18.4491	16.8910	15.5755	14.4502	13.9464	13.4765	13.0373
53	20.4924	18.5878	17.0072	15.6743	14.5352	14.0255	13.5504	13.1064
54	20.6552	18.7217	17.1192	15.7694	14.6169	14.1016	13.6214	13.1728
55	20.8125	18.8508	17.2271	15.8609	14.6955	14.1747	13.6896	13.2366
56	20.9645	18.9754	17.3311	15.9490	14.7711	14.2451	13.7552	13.2979
57	21.1112	19.0956	17.4213	16.0338	14.8438	14.3127	13.8182	13.3568
58	21.2529	19.2114	17.5278	16.1154	14.9137	14.3777	13.8788	13.4134
59	21.3898	19.3232	17.6207	16.1940	14.9810	14.4402	13.9370	13.4678
60	21.5219	19.4309	17.7103	16.2696	15.0457	14.5002	13.9930	13.5200

Dual rate leasehold valuation tables
(From Parry's Valuation and Investment Tables 11th Edition)
The College of Estate Management. Reproduced by kind permission.

only buyer in the market. This form of top slice profit rent would almost certainly be valued on a much lower years' purchase than the rent from a traditional leasehold.

For the mathematically minded, the years' purchase calculations for traditional leaseholds can be expressed in terms of relatively simple formulae. Take a freehold first. The number of years' purchase at which it is valued is obviously the reciprocal of the yield (if the yield is expressed as a decimal rate of interest). Thus, for a 5% yield the sum is:

$$\frac{1}{0.05} = 20 \text{ years' purchase}$$

For a leasehold, the sinking fund payment has to be taken into account. So, for our 50-year lease with a sinking fund at 4%, the years' purchase is the reciprocal of the yield plus the sinking fund payment. For a nil tax payer the sum is thus:

$$\frac{1}{0.05 + 0.0065502} = 17.6834 \text{ years' purchase.}$$

If the sinking fund payment has to be made out of income that has borne tax, there is an additional layer of complexity. If "T" represents the tax rate, the sum becomes:

$$\frac{1}{0.05 + 0.0065502\left(\dfrac{100}{100 - T}\right)}$$

With a 35% tax rate, the sum is thus:

$$\frac{1}{0.05 + 0.0065502\left(\dfrac{100}{100 - 35}\right)} = 16.645 \text{ years' purchase}$$

This is probably enough of the arithmetic. Most people will be content to use the figures already worked out for them in the valuation tables, though programmed calculators and computer programmes are increasingly supplanting the printed figures. But one point needs making. The valuation tables used to be calculated on the convention that rents were paid annually in arrear. As noted, this is the convention we have adopted for simplicity to illustrate the principles. In practice, rents are normally paid quarterly in advance, and calculating the sums on this basis gives a slightly different (and more realistic) answer. Modern valuation tables or calculator and computer programmes will provide the "quarterly in advance" option.

8
What land is worth

What is a piece of land worth? The obvious answer is that it is worth what you can sell it for. But how do you calculate what price a piece of land might be expected to fetch? How, indeed, do you calculate what you can afford to pay for a piece of land if you are the prospective buyer? There is, let us assume, no building on the land at present, nor is it producing any rental income. Since property valuation is frequently a matter of putting a capital value on a future flow of rental income, where do you start?

The answer is that the value of a piece of land depends on what can be done with it. It could have one value if its use was restricted to farming, a totally different value if you were allowed to build houses on it and a different value yet again if you had permission to erect an office block on the land (assuming there would be a demand for offices in that location). In other words, to see what a piece of land might be worth you have to visualise it in its developed state, then work backwards. This is what is known as a "residual" basis of valuation.

Take a simple example. On the piece of land in question — assume that it is freehold — you have planning permission to build an office block, for which you are confident that there will be occupier demand. So you begin by adding up the costs of erecting your office block. There will be the construction cost itself and the fees to architects and other professional advisers. Then there will be the finance cost: the interest on borrowed money while the block is under construction and before it is producing any revenue. You may also, depending on the economic and property market climate, want to build in a significant sum for marketing the block once it is completed and a further amount to cover the period for which you expect it to be empty; when the block is empty and producing no revenue it is still clocking up interest charges, which add to the overall financing cost.

Let us suppose that all these costs add up to £10m. But that is not the end of the story, because you are taking a risk if you decide to develop an office block and there is also no point in doing it unless you make a profit at the end of the day. So you build in an extra 20%, say, as your profit margin and to compensate you for risk. On top of the £10m of costs, that brings the total to £12m.

Then you look at the project from the other angle: what will it be worth on completion? In the prevailing climate you estimate that you should be able to let the offices on full repairing and insuring terms at a rent of £1m a year. Investors are currently prepared to buy properties of this type to show themselves a yield of 7%. This is the same thing as saying that they will be prepared to buy at about 14.3 years' purchase of the rent. So, if somebody is prepared to buy the completed building at 14.3 times the £lm rent which you expect it to produce, it will be worth £14.3m once it is completed and let.

Thus if your total costs — plus your profit allowance — are expected to be £12m, you could apparently, afford to pay £2.3m for the land (the £14.3m end value, less the costs and profit allowance of £12m). In practice it is not quite so simple as this, since you will also want a profit on the money allocated to purchase the land and the finance costs of holding the land for the development period. Your total costs include the land cost, after all. If you allow for this, you find you could pay only about £1.6m for the land. The sums then look something like this:

	£m
Land cost	1.6
Finance charges on land cost	0.3
Cost of construction, fees, finance, etc	10.0
	11.9
Add 20% for risk/profit	2.4
Total	14.3
Projected value of land and building	14.3

So far, so good. But what if your projections turn out to have been wrong? The answer helps to explain why land prices generally fluctuate far more wildly than property prices as boom and slump succeed each other in the property market.

Suppose, first, that the rental market improves significantly while your development is under way. Instead of achieving a rent of £lm, you let the building at £1.1m. Assuming it is still valued on a 7% yield — at 14.3 years' purchase of the rent — it will be worth £15.7m instead of £14.3m. You will have made a profit of £3.8m instead of £2.4m. Looked at another way, you could (with the benefit of hindsight) have afforded to pay as much as £2.6m (after allowing for finance charges) for the land and still have made your projected 20% profit.

This is what frequently happens in the early stages of a cyclical property boom. Rents and values move ahead fast so that, when it is completed in two years, a development will turn out to have been considerably more profitable than envisaged at the outset.

The danger comes as the boom rises towards its peak. With recent evidence of rapidly rising rents and values, developers tend to assume that the process will continue. So they build into their calculations an allowance for a rise in rents and values during the development period. Or, which comes to much the same thing, they settle for a lower profit margin based on the rents and values prevailing when they do their sums. They hope that rental inflation will assure that the ultimate profit margin is higher than budgeted at the outset.

Suppose, at this later phase in the development cycle, that a developer planned a project similar to our earlier example. The current rental value of his projected building would be £1.1m, but he reckons that this will have risen to £1.2m by the time it is ready for letting. On a 7% yield, that will be worth roughly £17.6m. He is still looking for a 20% profit margin and calculates that he can afford to pay £3.9m for the land (we have assumed, perhaps unrealistically, that construction costs have not increased). His sums then look like this (the figures are rounded for convenience):

	£m
Land cost	3.9
Finance charges on land cost	0.8
Cost of construction, fees, finance, etc	10.0
	14.7
Add 20%, for risk/profit	2.9
Total	17.6
Projected value of land and building	17.6

But he has got his timing wrong. By the time the building is ready for letting a massive surplus of newly developed office space has built up. The most he could hope to get in rent is £900,000 instead of his hoped-for £1.2m and, to make matters worse, investors are demanding a yield of at least 9% on the now less-popular office investments. On a 9% yield — at 11.1 years' purchase — the £900,000 rent gives the building a value of only about £10m, or considerably less than the construction and finance costs. So there is no profit for the developer and, with the benefit of hindsight, he should not have paid anything for the land. In fact, he should not have undertaken the development at all. Since building on the land would create losses, not profits, even if the land came free, in effect the land now has a negative value.

Cases like this were all too common in the depressed property market conditions of the early 1990s. Because land is the residual element in development calculations, small fluctuations in the values of completed properties can cause far greater changes in the value of land.

9
Measuring property performance

Far more information on property performance is now available than was the case 20 years ago, and it is possible to get a reasonable picture of the way the market is moving. As we saw in Chapters 3 and 4, there are three main factors affecting values of investment properties:

- A rise or fall in the rental value of the building.
- A rise or fall in the yield basis on which the property is valued.
- The approach of rent reviews or reversions.

Sometimes other factors will also intrude. For example, if a profitable opportunity occurs to redevelop an existing building and replace it with a building that would bring in a higher rent, the value of the existing building will rise to reflect this development potential.

But before looking at the common measures of performance for the property market, there are a few caveats. An index of property performance cannot be regarded in quite the same way as an index of shares: the *FTSE* Actuaries All-Share Index, for example. The All-Share Index is based on the actual stock market price of the constituent shares. The current price for a property is generally a valuer's best estimate and inevitably contains an element of opinion.

No two buildings are exactly alike and no index of property performance should be taken as a measure of what is likely to be happening to the value of a particular building at a particular time. Office buildings in the City of London are probably moving in value at a different pace (and sometimes even in a different direction) from offices in Manchester, and a modern building in Manchester may not be rising or falling in value at the same rate as an older one.

Nowadays a property index may be broken down into a number of sub-indexes, by type of property and by geographic location. But it is still best used as a measure of the trend and the mood of the market in a particular area, rather than as an absolute yardstick against which a particular property portfolio is judged.

Property indexes can be constructed in two main ways. In one method, a portfolio of actual properties is monitored to provide a yardstick for the market as a whole — the proportions of different classes of property will probably be selected to mirror the composition of a typical institutional portfolio. A variation on this approach is to base the performance data on information supplied by a large sample of institutional property investors. The second approach is to construct a hypothetical portfolio with "rent points" at representative locations across the country and an estimate is made of what is happening to rents (and, perhaps, yields) at these locations.

In the case of an index based on actual properties in a portfolio, four main indicators of performance may be provided. First is a measure of the "estimated rental value" or ERV of the individual properties. This is the rent they would produce if let in the open market today.

Second is the actual rent that the properties are producing. This will not always be the same as the estimated rental value because, if rents are rising, the ERV will reflect the movement but the actual rent from the property will not increase until the rent review occurs. So movements in rental income will lag behind movements in the ERV when rents are rising.

Third is the capital value of the building. This will normally rise to reflect an increase in the ERV, but at a time when rental growth is accelerating it may not rise quite so fast because of the discounting calculation referred to in Chapter 4.

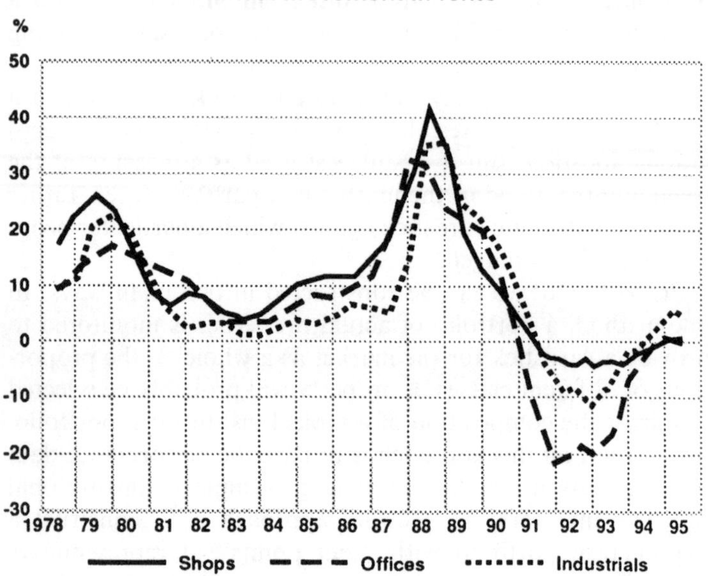

The graph, which shows the year-on-year percentage movement in rental levels for the three main classes of property, derived from Hillier Parker indexes, emphasises the volatility of the office sector. Rental levels for offices fell far more sharply in the early 1990s than those for shops or industrials. But note that "rental levels" is the operative phrase. Thanks to upward-only rent review clauses in leases, few existing tenants benefited from the falling rental levels, giving rise to the phenomenon of over-rented properties. Source: *Hillier Parker*

Finally, there is the overall return to the investor. This is a combination of the income he receives from rents and of the movement in the capital value of the buildings.

From the investor's viewpoint the overall return is the most important figure, but he will want to see how it was achieved. Capital values might have risen simply because yields had generally moved down which, as we have seen, is less satisfactory in the long run than growth based on increasing rental values.

Likewise, the investor will want to see what types of property were performing best. An index embracing all types of business property may be produced. But the breakdown of the separate performances of the main classes of property — shops, offices, industrial buildings and, perhaps, agricultural property — may be more useful, particularly if there is a further breakdown by geographic region.

There may also be an indication of the contribution to capital growth from the different factors we have identified. If values have risen overall by 10%, perhaps 7 percentage points of the increase was accounted for by rising rental values, 2 percentage points came from a slight downward movement in yields and 1 percentage point came from the approach of rent reviews and reversions. Even if this breakdown is not given, it is usually possible to get a rough picture of what has been happening from the other figures provided.

The second type of index, based on "rent points", by itself simply provides a sensitive index of the way rents are moving. But it is possible also to estimate the yields on which buildings at the rent points would be valued and from the combination

INDEX VALUES

All properties

	Rental value	Capital value	Total return	RPI
87 Dec	115.24	110.16	117.05	103.7
88 Dec	142.10	136.67	153.62	110.7
89 Dec	165.17	151.34	179.29	119.2
90 Dec	173.45	133.31	167.81	130.3
91 Dec	166.60	122.81	166.56	136.1
92 Dec	150.60	112.71	166.45	139.6
93 Dec	138.61	119.83	193.70	142.3
94 Dec	136.44	127.62	223.25	146.4
95 Jun	135.82	124.31	226.19	150.2
95 Jul	136.14	123.69	226.57	149.5
95 Aug	136.09	123.38	227.50	150.3
95 Sep	136.04	122.99	228.30	151.0
95 Oct	136.20	122.62	229.13	150.2
95 Nov	136.19	122.12	229.74	150.2
95 Dec	136.04	121.61	230.33	151.1
96 Jan	136.13	121.50	231.68	150.6
96 Feb	136.30	121.34	232.93	151.3
96 Mar	136.53	121.28	234.39	151.9
96 Apr	136.72	121.15	235.73	153.0
96 May	136.84	120.95	236.94	153.3
96 Jun	137.05	121.34	239.33	153.4

How Investment Property Databank measures property performance of rental value and yield an index of capital values and overall returns can be produced.

Note, however, that performance figures produced in this way from a "rent point" type of index are to an extent theoretical and are not designed to mirror the likely performance of an actual portfolio. In a real portfolio the location of some properties is improving while other properties are being downgraded in investment appeal. With the "rent point" approach, the prime location in the particular town is normally selected, and if fashion shifts to another area the rent point will move to reflect the change. In addition, the calculations make no allowance for the fact that the landlord will need to wait for a rent review to collect any increases in income. On the other hand, the "rent point" approach probably gives the earliest indication of the way rents and values are moving in a particular area.

Other indicators are produced which simply measure shifts in the yield basis of valuation. In the past these were confined to the prime yield: the yield on which the property with the very best combination of growth, low risk and so on is valued. Nowadays indexes are produced which reflect shifts in the average yields on which property likely to be found in an institutional portfolio might be valued. The yield in question is, of course, the equivalent yield that a valuer would use, not necessarily the yield that the property owner is actually receiving on the current capital value of his properties. Rent indexes for secondary properties are also available.

The important point is: if you make use of an index of the property market, make sure you know how it is constructed and what it is designed to show.

10
When valuations are needed

Valuation of property, we frequently hear, is an art not a science. As such it must be one of the few art forms — apart from "creative accounting" — to surface repeatedly in company prospectuses, company accounts and a whole range of other legal and financial documentation.

We are concerned here not so much with the technical details of valuation as with the way it is used. Why do valuations matter? And what are they designed to show?

As a starting point, take the values of properties in company accounts. The shares of most industrial and commercial companies quoted on the stock market are valued mainly on the criterion of the earnings the company produces. But companies whose main business is owning or developing properties — property companies of various types — are rather different. The value of the shares may be expressed in relation to the value of the assets the company owns. "Payola Properties at 345p is standing at a 20% discount to net assets", and so on.

If a company is valued mainly in relation to its assets, the value put on those assets is clearly important. And in practice quoted companies are required to observe a mix of rules on property valuations: accountancy recommendations, Companies Act requirements, Stock Exchange regulations and guidelines from the Royal Institution of Chartered Surveyors (RICS). When bids are in the air, the City Takeover Panel also enters the picture.

With this welter of rules and advice, surely there should be a standard way of treating property values? Unfortunately, no. Some companies want to play up their worth. Some want to play it down. And others, for legitimate (or sometimes less legitimate) reasons, would rather keep investors and others guessing.

So at the outset there is one accountancy point to deal with, because it can provide a significant loophole for companies that like to surround their real worth with a convenient layer of fog. It will crop up again when we come to examine property company accounts in detail.

The problem is that properties can appear at two different points in a company's accounts. They may be shown as a fixed asset of the company, alongside any plant and equipment the company owns. Or they may appear under current assets, alongside stocks of materials and other items that are being turned over rapidly in the business. A company that sees itself as a long-term holder of properties — as an occupier or investor — is likely to show them as fixed assets. A trader in properties will probably show the real estate it owns as a current asset — part of its stock-in-trade. The distinction can affect the tax and accounting treatment. It certainly affects the requirements for valuation.

Current assets are normally shown in the accounts at "the lower of cost or net realisable value". In other words, the properties will appear at cost unless they are reckoned to be worth less than this figure. To complicate matters, sometimes they are shown at cost plus a proportion of the interest charges on borrowings incurred to buy them. The important point is that the figure in the accounts for properties shown as current assets will usually give you very little idea of what they are really worth.

Often there are logical and legitimate reasons for holding properties as current assets. But the distinction between a property that is a fixed asset and one that is a current asset can be pretty arbitrary. A company which wishes to remain vague, for good or bad reasons, as to the value of its properties may find a way of classifying them as current assets.

Properties held as fixed assets are a different matter. They may be shown at cost, but if their value is sufficiently different "in the directors' opinion" to be of importance to shareholders, the Companies Act requires an indication of the difference "with such degree of precision as is practicable". There is still plenty of room for opinion. But at least the accounts (or the notes to them) must break property holdings down between freeholds and long and short leaseholds, though details of individual properties are not required to be published. And where properties have been included in the accounts at revalued figures, the date of valuation and nature of the valuer must be stated. These provisions apply to all companies, not only property companies.

The accountancy bodies have been looking recently at proposals which could require periodic revaluation of current-asset properties (see Chapter 16). But until such proposals are implemented — if they are — the distinction between properties held as fixed assets and those held as current assets remains to be exploited.

When we come to property companies, additional considerations apply. The relevant accounting standard (*SSAP 19*)

recommends that investment properties — not properties held as stock-in-trade, note — should be valued at least every five years by an external (i.e., independent) valuer and shown in the accounts at open market value, though individual details are not required. And they should be valued every year by a qualified professional (though in the intervening years he might be an in-house valuer).

However, this is still less stringent than the rules imposed by the Stock Exchange on companies coming to the market for the first time. The Exchange's listing requirements have a special section on property companies and include a specimen valuation report showing how the details of properties should be set out.

Broadly, companies are required to provide details of each individual property (though some lumping together may be allowed when properties are too numerous). Valuation by an external valuer, in accordance with RICS guidelines, is normally required, though an internal valuation might be allowed if the company has a "qualified surveyor's department". Quite a lot of detail has to be given on each property, including tenure, main terms of lease, capital value in existing state, and so on. Where holdings of land for development or actual developments are included, the valuation report must say whether planning permission has been obtained, when the development is expected to be completed, what it will cost to complete, open market value in its existing state and estimated capital value when completed and let. This is usually the first and last time shareholders get information in such detail. And the requirements normally apply to all properties, whether held as fixed or current assets, except for a possible let-out which we will come to later.

If a property company, once listed on the Stock Exchange, makes a major acquisition of further properties that qualifies as a "Super Class 1" transaction (broadly, one that adds 25% or more to its size) it will need to provide a valuation of the properties being acquired.

When a company buys from (or sells to) its own directors, substantial shareholders or their associates — this used to be known as a "Class 4" transaction in the Stock Exchange terminology, but is now referred to as a "related party" transaction — an independent valuation will be required. The reasons for this provision do not need spelling out.

And when defending itself against a bid, a listed company *may* be justified in publishing not only a valuation of its completed properties and developments in their existing state, but also an estimate of prospective values of developments, provided the assumptions are stated.

The City Takeover Code has more to say on the subject of bids. "When a valuation of assets is given in connection with an offer, it should be supported by the opinion of a named independent valuer." Investment properties should be valued at open market value. If assumptions are made in a valuation, they must be spelled out and the Panel must approve them. In the case of developments, the requirements are similar to those imposed by the Stock Exchange, but the emphasis is different: costs and estimated eventual values *should* be stated.

However, there is a possible concession to property companies at the receiving end of a bid and with insufficient time to get a valuation completed before the defence document goes out. "Exceptionally" the Panel might allow informal valuations to appear. And in other "exceptional" circumstances a valuation of a representative selection of the properties might be allowed in place of a full valuation.

One other document that may provide considerable detail on a company's properties is the prospectus issued when the company issues a secured loan — such as a mortgage debenture stock — to investors at large. However, the properties covered will be only those specifically charged as security for the loan. If these are only a proportion of the group's properties, details of the remainder will not be provided.

We will see in later chapters what these varying requirements mean in practice — and some of the ways in which the sharper operators may circumvent them.

11
How accurate are valuations?

We looked in the last chapter at the main requirements from different bodies on the valuation of properties in company — and particularly quoted property company — accounts. To summarise:

- The most comprehensive valuations of the properties that a property company owns are required when it launches on the stock market, when it uses an "asset value defence" in a take-over bid or when it publishes a valuation in connection with a mortgage debenture issue.
- Ongoing valuation requirements are far less rigorous. The accounting standard recommends an independent valuation of investment properties at least every five years. The Companies Act simply requires an estimate of the difference between book value and market value when this is significant.
- If a property company chooses to classify its properties as trading stock rather than fixed assets they will normally be shown at cost and most of the requirements to provide a valuation do not apply — except possibly in a stock market launch.
- The treatment of properties under development is problematic. Companies will normally show them at cost until completed and let. But in a market launch or an asset-based bid defence they will be required to estimate eventual value and the costs involved in completing them.
- Requirements as to methods of valuation are normally expressed in terms of the *Guidance Notes* of the Royal Institution of Chartered Surveyors (RICS).

There are a couple of other points we have not yet mentioned. In some contexts, when property companies produce up-to-date valuations or estimates of values, they are also required to estimate the tax that would be payable if the properties were sold at these values — this would be needed in order to calculate a "break-up" net asset value for the company.

And property dealing companies may be treated rather differently from property investment and development companies even under the Stock Exchange's listing requirements. Dealing companies would normally hold their properties as current assets and they tend to argue that they are valued by the stock market on the profits they can generate rather than the assets they own, which they are constantly turning over and which are therefore less relevant. It is a distinction that can be exploited in some circumstances. The Stock Exchange examines each case on its merits, but will not always insist on a full valuation for a dealer.

So what should we expect to learn from a property company's accounts about the properties it owns? If they are held as fixed assets we can see whether they are freehold or long or short leasehold, whether they are shown at cost or valuation and, in the latter case, when they were valued and by whom. If they are shown as current assets, we may know what they cost (with or without unquantified financing charges) and very little else. Some companies may choose to give additional useful information. Others will not.

Property companies traded on the stock market span a range of valuation practices. Some produce an independent valuation each year. Others choose to have a proportion of their portfolio independently valued each year so that, say, each completed property will have been independently valued at least once every three or five years. In the interim the directors may produce an estimate of any surplus on parts of the portfolio that were not revalued that year. Others carry out a full valuation each year by professionally qualified directors and incorporate the results in the accounts, resorting to independent valuers only once every five years.

Some companies deliberately adopt the most conservative basis of valuation, reckoning that they do not want to disappoint shareholders if conditions in the property market deteriorate. Others — particularly those who would like their shares to stand at a high price so that they can be used as takeover currency — may prefer to put everything possible in the shop window, even providing estimates of likely surpluses on completion of developments.

But many of the more established companies like to have something in reserve. If an unwelcome bidder looms on the stock market horizon, it is remarkable how much additional value a company may sometimes uncover in its properties as it attempts to make itself too expensive for the predator or to exact the best possible price. Part of this additional value may come in the form of expected surpluses over cost on the developments it is undertaking.

And how accurate is a valuation anyway as an estimate of what a property will fetch? There are optimistic valuers and

pessimistic valuers. There are property owners who encourage their valuers to be optimistic and those who encourage them to be pessimistic. Optimistic valuers may turn out to be right in a fast-rising market, pessimistic valuers in a dull market. And then there are "special assumptions" — the subject of much hot debate in the surveying profession.

The RICS *Guidance Notes* on valuation focus on "open market value" in the case of properties held as investments — what a willing seller might expect to have obtained given reasonable time to negotiate the sale and free exposure to the market. The concept specifically excludes consideration of what a buyer with a special interest might pay.

But what might the property be worth to, say, a buyer who owns an adjoining building and could combine the two in a single development that would be more valuable than the sum of the parts? Or — a similar argument — should a large portfolio of properties in any case be valued at a figure higher than the sum of its parts, since considerable costs would be incurred in assembling such a portfolio from scratch?

The second argument in particular crops up in the context of takeover bids for property companies, where figures paid often seem high relative to those thrown up by conventional valuation. The RICS has stuck to its guns in defending "open market value" as the normal basis for valuation. But where "special assumptions" might result in a higher figure, it allows this to be pointed out. However, the assumptions will in future have to be spelled out.

How safe are we even with conventional valuations? A few years ago an enterprising actuary, David Hager, decided to put it to the test. He selected two investment properties and got an initial "control" valuation from one valuer. He then let loose on the properties nine other valuers. In one case their figures ranged from 13% below the control valuation to 7% above. In the other the range was 25% below to 8% above.

Critics of the exercise claim that not all the valuers were familiar with the particular markets and that they were, in any case, asked to do back-of-the-envelope valuations.

Perhaps it is fortunate all the same that the price of a property company's shares is ultimately determined by the stock market rather than the valuer's apparently imprecise art. But there are other types of property-related investment where the asking price depends entirely on the valuer's opinion. These cases, and some of the other financial byways of the valuation process, are examined in the next chapter.

12 Valuing from the armchair

There has always been considerable reluctance among financial regulators to allow the sale to the public of investment products whose price depends entirely on a property valuation. We have already seen how different valuers can come up with widely different figures for the same building, so the caution is perhaps understandable. There is also a fear in the minds of the regulators that property is too illiquid an investment for the man in the street — in other words it may be difficult for him to get his money back if this involves trying to sell properties in an unreceptive market. At best, property sales take time.

So, in the past, authorised unit trusts (those whose units are on sale to the general public) were not allowed to hold property direct as an investment. They could buy shares in property companies, since these could be traded in the stock market like any other kind of share, but they could not buy a chunk of real estate. That has now changed. Authorised unit trusts are allowed to own property direct. But the climate of the early 1990s was hardly conducive to property-owning unit trusts and initially there were few examples.

But even before this change, there were forms of investment whose value was linked directly to the value of a property portfolio. There were (and are) the unauthorised property unit trusts for pension funds and charities; these are not open to the general public. And there are property bonds — technically life assurance contracts whose value depends on the value of a portfolio of properties. In the latter case the investor buys a certain number of units, gets a minimal amount of life assurance cover and watches the value of his units rise or fall with the movement in the values of the properties in which the "life fund" invests. The technicalities and the tax treatment are different, but in practical terms it is not dissimilar to the unit trust principle.

This is where valuation comes into the picture. We have seen that a property company may need to revalue its investment properties, and that the stock market price of the property company's shares will be influenced by the asset value and may be expressed in asset value terms ("stand at a premium or discount of such and such an amount"). But what determines the share price at the end of the day is what investors will pay for the shares in the market. It is not directly linked to the asset value.

The unit trust or the property bond is different. The managers have the portfolio of properties valued, make whatever adjustments are needed, then divide the total value by the number of units in issue. And, bingo, that is the value of a unit. Or, more accurately, it is the value at which managers will sell further units or buy back existing ones. (In practice they produce two prices, a lower one at which they will buy and a higher one at which they will sell, the difference being accounted for by various costs and charges.) So the price is not directly determined by a market but by the managers, with the help of the valuation profession.

If we remember that the difference between a high and a low valuation might be as much as 25%, the dangers hardly need spelling out. The investor who bought units at a price determined by a high valuation, and found when he came to sell that the valuer was taking a pessimistic view instead, could lose very heavily.

There is another problem. How frequently should the properties be valued? Once a year is clearly inadequate — the unit price would stay fixed for a year, during which property values might have been soaring or slumping. But full property valuation can be an expensive business. Full monthly valuations could absorb most, if not all, of the return from the properties.

Enter at this point what used to be called the "armchair valuation". But as that sounds a little too relaxed, valuers nowadays prefer the term "desktop valuation". The meaning is the same. A "valuation" is undertaken without even looking at the buildings in question.

This is not so bizarre as it might appear. If, following the last full valuation, you have all the details of a property on computer — size, rent, lease terms, review date, rental levels in the area, yield applied in the original valuation, and so on — it is comparatively easy to adjust for changes in any of these factors to produce an updated valuation figure. And since the process is a lot less expensive than a full valuation, it can be undertaken fairly frequently. Property bond managers rely heavily on desktop valuations to adjust the unit price each month.

There are a few other oddities of the valuation process that the investor in property will come across. Not least is the question of "marriage values", which impinge on the area of "special assumptions" in valuations that we have already looked at.

Take a simple example. A property developer put up and let an office building in the late 1950s, before inflation and shortage had caused rents to begin their rapid climb of the post-war era. He thought he was being clever in finding a good household-name tenant who would take a lease of the whole building for 99 years — with only one rent review after 50 years. The valuation of the building after the original letting showed him a good surplus over his costs and he was assured of a regular income virtually into perpetuity.

By the late 1980s the transaction was looking very different. The rent being paid by the tenant was now a mere fraction of the market level. The owner had a fixed-interest investment for another 20 years and the reversion to a market rent was still so far off that its discounted value was comparatively low.

The tenant meanwhile holds a valuable asset, though ultimately a wasting one. But the right to occupy a building for 20 years or so at a fraction of its open market rent is not to be sneezed at. It is also a saleable commodity.

The point is that the value of the asset in the hands of the owner and the value of the asset in the hands of the tenant would add up to less than the total value of the building if it could be let today on a modern lease at a market rent. The difference is the potential "marriage value" of bringing the two interests together.

The tenant might offer to buy the landlord's interest. The landlord might offer to buy out the remainder of the tenant's lease. The landlord's interest is worth more to the tenant than to anyone else and the tenant's interest is worth more to the landlord than to anyone else because, by combining the two, they can come up with something that is worth more than the sum of its parts. In practice, a deal is normally struck in which the marriage value is divided between the two parties, regardless of which one buys the other out.

And even if, say, a third party were to buy the landlord's interest instead, he might reckon that it was worth paying more than the apparent value because of the possibility of realising the marriage value by doing a deal with the tenant at some future point. In these circumstances the RICS *Guidance Notes* allow the "hope value" to be included in an open market valuation. Ultimately, the tenant can afford to pay a higher price than anyone else. But if everybody in the market is likely to bid a price that reflects the hope of doing a deal with the tenant, this price becomes the general view of the market.

13
Property financing constraints

Having looked at some of the investment and valuation aspects of property itself, we now need to see how these are reflected in one of the most common vehicles for owning property: the property company.

But before we get into the detail of property company structures and property financing techniques, a few of the financial characteristics of property are worth examining. We are talking here about "equity" investment in property — investing for growth as a risk investment. (There are other property-related forms of investment such as mortgages or ground rents which may be merely a fixed-interest investment.)

Investing for growth usually implies accepting a low income return from property today in return for the expectation of increasing income and capital values in the future. But this poses a problem in the short term. If a property shows an immediate yield of, say, 6%, you will have difficulties if you want to borrow the money to buy it. Rates of interest will vary according to the type of borrowing and may change from one month to the next. But let us assume for convenience that borrowed money costs you 10% a year.

The point is that the rental income which the property produces is not enough to cover the cost of the interest on the money you borrowed to buy it. Say you borrow £1m to buy a property and it shows you £60,000 a year initial income. The interest on the debt costs £100,000 a year. Expenditure exceeds income. Disaster.

This would not necessarily have been the case in the immediate post-war era, before inflation caught hold. Rates of interest on mortgages were lower than yields on property because mortgages were regarded as safe, whereas equity investment in property carried risk. It was only later that investors came to appreciate the growth possibilities in property. And realised that the greatest risk of all was that money lent by way of "safe" mortgages could have lost most of its value by the time it was repaid. In the property crash of 1990-93, yields on some properties again moved above borrowing costs, but this seemed likely to be a temporary phenomenon.

Today, even the investor whose expenditure exceeds his income may actually be making a positive overall return. The combination of income and capital growth from the property might give an overall return of 15% a year in a reasonably buoyant property market. This comfortably exceeds the 10% interest cost. But this is not a great help when you find yourself without the cash to meet the £100,000 interest bill. On paper, you are making a positive return, over your borrowing costs. But you have a negative cash flow.

This is a vital distinction. Many property financing techniques are designed to get round this problem. If your property is rising in value, the simplest answer might be to borrow additional money each year against the increasing value of your property, and use this cash to pay part of your interest bill. But even if this were always possible, there would be dangers. It is what many property companies were doing in the early 1970s and is the reason why many of them got into serious or terminal difficulties in the property and financial crash of 1974-76. Property values slumped. Property became virtually unsaleable and impossible to borrow against. And the results of "deficit financing" techniques — buying property with borrowed money that cost more than the rents yielded — became all too apparent. The companies' outgoings exceeded their income, they could not raise cash to cover the difference and they were in trouble.

Of course, if you can buy the property with your own cash and you have no great need for immediate income, the problem does not arise. You only get a 6% a year income return initially. But when you come to sell the building at a profit at the end of the day, your overall return will have been 15% a year. This is why institutions such as pension funds and insurance companies are natural long-term property investors. They use their own money (or that of their savers). They do not use borrowed money, and they do not therefore have to worry about interest charges which exceed the rental income.

For most property companies the situation is different. They use a mix of their own money ("shareholders' funds" or "equity finance") and borrowed money. In other words they are "geared" (or "leveraged" in the American terminology). "Gearing" is the relationship between borrowed money and your own money — the equity — in a business.

Gearing has advantages for property companies when rents and values are rising. Say a company borrows £1m at 10% to meet half the cost of buying a £2m property showing an income of £120,000 a year (a yield of 6%). Its own "equity" in the building is the £1m it put up itself. Say that rents are rising very rapidly and that after five years the rent doubles and the capital value of the building also doubles.

The company still only owes £1m but the building is worth £4m, so the value of the company's equity in the building has risen from £1m to £3m. In other words, thanks to the gearing the value of the company's interest in the property has risen 200% for a rise of only 100% in the value of the property itself. And whereas initially interest charges took £100,000 out of a rental income of £120,000, leaving the company with a return of only £20,000 on the £1m of its own money it had put in, the rent has now doubled to £240,000, the interest charge is

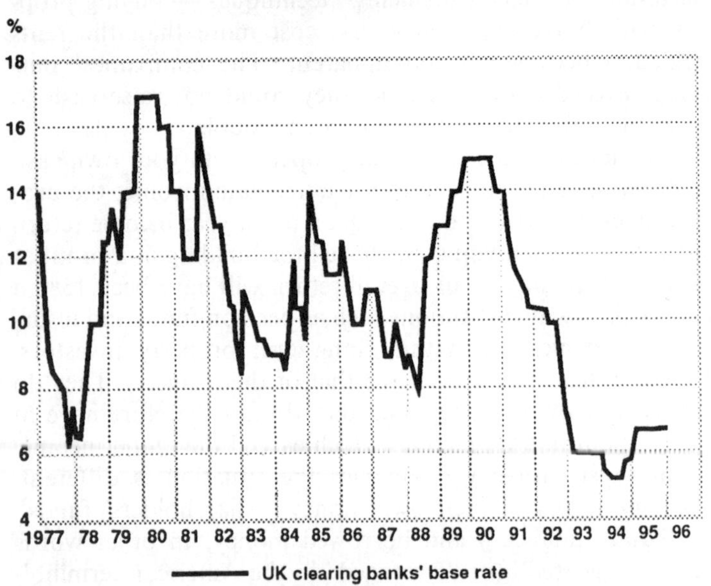

Bank base rates give an indication of short-term borrowing costs, though remember that even the safest borrower will normally pay a little more than base rate. The wild gyrations in interest rates over the past couple of decades underline the problems that property developers and investors have to contend with. When base rates roughly doubled in the late 1980s, developers who had failed to "cap" their borrowing costs sometimes found themselves clocking up interest on their development loans at double the rate they had expected. It was a major factor in the subsequent property market collapse. But the interest rate climate was looking considerably calmer by the mid-1990s. Source: *Datastream*

still only £100,000, so the income the company keeps rises from £20,000 to £140,000 a year — a rather more realistic return on £1m.

Of course, gearing works in reverse, too. If property values are falling, your own equity loses value faster if it is geared up. Much of the art lies in judging how much gearing is safe, and when.

In the example we have just looked at, the company borrowed only half the cost of buying the property. And, in practice, lenders do not like putting up the full cost. The borrower must normally put up an equity "cushion" of his own money. But an established company owning revenue-producing properties could use these as security for a loan to cover the full cost of buying an additional property. If the interest on the borrowing exceeds the income from the new property, the shortfall could be made good from the income from the other properties. And with luck, at the first rent review the income from the property will rise to cover the borrowing costs. Established companies with a good income stream have an advantage.

When we come to property development there is an additional problem. Suppose it takes four years from inception to complete an office development. The developer has to find the money to buy the land and to fund the construction work. And he has to pay interest on this borrowed money. Yet his development is just a building site — it is not producing any income at all. So, either he has to meet the interest costs from his other revenue or he must "roll them up" in one way or another. "Rolling up" means they are added to the amount of the loan that has to be repaid at the end of the day. In either case, finance costs are an integral and important part of the cost of any development.

The payoff is that the development will, if all goes well, be worth more at the end of the day than its cost — though this is not necessarily the end of the financial problems.

14

Investors and traders

When a developer wants to keep his buildings as an investment once they are completed and let, the real financing problems begin. Take even an (exceptionally successful) office development costing £10m and estimated to produce a rent of £900,000 a year at the end of the day. The £10m cost, remember, includes the finance charges incurred during development.

If the building is completed on time and finds a tenant who will pay £900,000 a year, the offices would be worth just over £15m to an investor on a 6% yield or at 16.7 years' purchase . This is a surplus of £5m over cost. The developer could simply sell the building at this point and pocket his £5m profit (less tax).

But look at the position if he wants to keep the building (assume for purposes of illustration that he has borrowed the whole £10m cost). The interest charges on the £10m borrowings are, say, £1m a year (and in practice probably quite a bit more), but his income from the building is only £900,000. He may reckon that it is worth holding on and trying to find a way of meeting the £100,000 a year income shortfall. If he can hold on till the first rent review in five years, the income should rise to exceed the interest cost and he will also have an additional capital profit on the building.

But meantime interest charges have to be met and so, for a property company, life becomes a constant juggling between the need for income and the desire for long-term capital gain.

It follows that property companies fall into two main categories. Those that do try to hold the properties they buy or develop as long-term investments for a rising income and appreciation in value. And those that develop properties for sale, content to make a trading profit and move on to the next project.

Thus there are also different concepts of "profit" in the property company world. Let us go back to the example of an office development costing £10m, but assume this time that half of the cost (£5m) was borrowed at 10% interest and the other £5m came from the company's own resources. The building is let at £900,000 a year rent and is valued at £15m. Assume the development company owns only the one property. At the end of the first year after completion and letting, the company's profits could look like this:

	£
INCOME: (rent)	900,000
INTEREST PAYABLE:	500,000
PROFIT BEFORE TAX:	400,000
CORPORATION TAX:	140,000
NET INCOME:	260,000

or they could look like this:

PROFIT: (on sale of property)	5,000,000
INTEREST RECEIVABLE (on company's own cash)	500,000
PROFIT BEFORE TAX:	5,500,000
CORPORATION TAX	1,925,000
NET PROFIT	3,575,000

Remember, these are two totally different but equally acceptable ways of dealing with the same set of circumstances. In the first example the developer holds on to his completed property as an investment. His property, valued at £15m, was worth £5m more than cost, so he has a surplus of £5m. But this does not appear in his income statement because he has not realised the profit by selling the property. So his income is simply the difference between the rent he receives and the interest he pays. This is subject to 35% corporation tax, assuming he has no allowances to claim (remember that for illustrative purposes we are sticking with the 35% corporation tax rate applying in 1990-91).

In the second example the developer regards himself as a trader. So he sells his completed development as soon as possible (we have assumed it was close to the beginning of his financial year), pays back the borrowed money and gets back the £5m he put into the development himself, on which he can then earn interest. His published profits are enormous compared with those in the first example. But he has nothing else. If he wants to show profits from property development the next year he must undertake another development and sell it on completion. His profits are likely to jump up and down, depending on his success, even though accounting practice allows some smoothing of the profit flow.

Now look again at the first example. The developer/investor has a comparatively modest net income of £260,000 a year

after tax and he still has £5m of his own money tied up in the company. That income is unlikely to change until the first rent review on the building in five years. But it is a very, very stable source of income.

Unless the tenant should go bust and the building become empty, the rental income will keep on rolling in year after year, almost certainly rising at five-yearly intervals in reasonably buoyant conditions for the property market. And the capital value of the building should be rising at the same time. Note that if you hold on to the building you do not pay any tax on your development profit or its subsequent increase in value — though tax would be payable if you sold it in the future. But if you sell the building on completion the tax man takes his bite right away.

Call the company that holds on to its building as an investment Investor Plc. Trader Plc is the one that sold the development on completion. How do we value them?

Investors are often very confused about how to value property companies. The stock market's main measure of companies in general — industrial companies, retailers and so on — is the price-earnings ratio or PE ratio. It is rather like the "years' purchase" applied to property. It tells you how many times you are paying for the company's present earnings (after-tax profits attributable to ordinary shareholders) if you buy the shares at their current price. If the company earns 10p per share and the share price is 120p, the PE ratio is 12 because at 120p you would be paying 12 times the current earnings of 10p. By and large, companies whose earnings are expected to grow rapidly will be on a high PE ratio, and those with more limited growth prospects will be on a low one.

Unfortunately, this does not help a great deal with property investment companies. Clearly, it would be inappropriate to value a property investment company in the same way as a manufacturer of widgets.

The problem is that an investment company's current income does not necessarily tell you very much about the value of its assets or the income they might be capable of earning in the future. A development financed mainly on

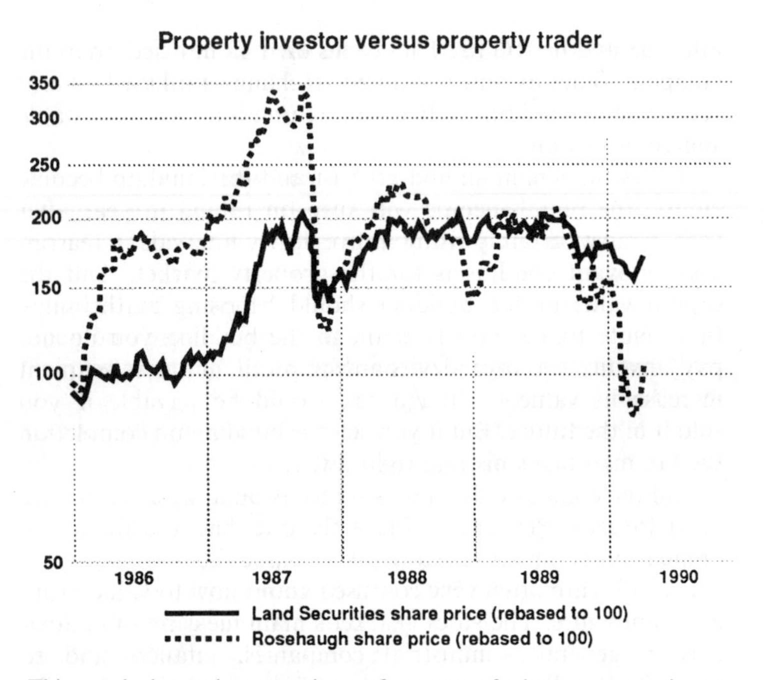

This graph shows the respective performance of a large property investment group and a large property trading group in different market conditions. The property trader, Rosehaugh, outperformed the investment group, Land Securities, in the share boom up to late 1987. But it was a different story when the property sector hit hard times and cash flow worries intervened at the end of the decade, with the investment company showing its defensive qualities and the trader seeing its share price fall heavily. With the benefit of hindsight, the market had it absolutely right. Rosehaugh later went bust. Land Securities survived the recession in reasonably good shape. Source: *Datastream*

borrowed money might initially show little if any income after interest charges, but it could be a valuable capital asset and the income could rise very sharply when rents are reviewed.

An example illustrates the point. If we apply a fairly typical (for an industrial company) PE ratio of 12 to Investor Plc's net income of £260,000, we find that the company would be worth £3.12m (£260,000 X 12). But the company owns a property worth £15m, subject to debt of £5m. If we knock off the £5m the company owes, its interest in the property is still worth £10m. Clearly it would be ridiculous to value a property company with net assets of £10m at only £3.12m.

Then take Trader Plc. If we apply the same PE ratio of 12 to Trader's profits of £3.58m after tax we find the company is worth £42.9m — more than 10 times what Investor Plc is worth on a PE ratio basis. Yet Investor Plc owns a valuable property. Trader Plc owns no property — it has sold its building — and simply has £5m of its own cash in the balance sheet plus whatever part of the profit on the sale of the building it has not paid out in tax and dividends. There's no guarantee that it will make profits as high as £3.58m next year — or, indeed, any profits at all.

As we have already hinted, what happens in practice is that the stock market judges Investor Plc mainly in relation to the net assets it owns and Trader Plc mainly in relation to the income it produces and its future income prospects. Let us try to quantify these figures.

In practice, Investor Plc might be worth around £8m in the stock market, on the assumption that it was valued at a 20% discount to the £10m of net assets. This discount varies according to the company and according to market conditions at the time — we have taken 20% as probably not being far off the long-term average. Early in 1990 discounts were actually closer to 40%, but the property and property share markets at that point were in particularly depressed mood. Early in 1994, some property investment company shares were actually standing at a premium to net assets, but this was mainly because the next revaluations were expected to throw up substantial surpluses.

What Trader Plc would be worth is problematic. If it had a good record of producing property trading profits year after year, investors might indeed award it a PE ratio around the average enjoyed by industrial and commercial companies (say, 12 — it will depend on market conditions at the time) giving it a market value of £42.9m. Other more cynical market watchers would reckon it should be worth the profits it can produce in a single year — plus whatever assets it has — and no more. On this basis it would be worth around £4m. In buoyant times the more optimistic view is likely to hold sway.

The precise figures do not matter too much. What is important is the principle. If you are buying shares in a pure property trader with little in the way of net assets, you are not investing in property. You are simply investing in the skill and ingenuity of a management that attempts to produce property trading profits each year. This is a very different matter from investing in a property investment company, where your shares are backed by property assets and by secure rental income, and where income and asset value will probably continue to grow even if the company undertakes no new activities at all.

It is very easy to produce property trading profits for a year or so by selling what an investment company would regard as the fixed assets of the business. The next trick is to persuade

investors to value your company at 10 or 12 times those (perhaps never to be repeated) trading profits. It is rather like the home-owner who says: "I bought my house at £40,000, built an extension for £10,000 and after a couple of years I find the house is worth £100,000. If I sell it at that price I make a £50,000 profit. On a PE ratio of 10 that makes me worth half a million". It is a nonsense, but the markets do not always spot it.

Thus property trading companies frequently try to acquire a high share rating on the basis of a short-term profit record, and use these highly-valued shares to launch a takeover bid for another property company which owns some solid assets. Unlike stock market investors, operators in the property market itself tend to focus on assets and generally regard trading profits mainly as a way of financing the purchase of assets.

The lessons from this are clear. The investor has to look carefully to see where property income comes from, and whether it is likely to be a continuing and growing source of revenue. We have talked so far in terms of a pure investment company or a pure trading company. But many property companies are in practice a mix of investor, developer and trader and each one needs examining on its individual merits. How much of the profit comes from rental income? How much from profit on property sales?

And this points to a further technique a developer may use to overcome the deficit financing problem that makes it difficult for him to hold on to buildings created with borrowed money. He may decide to hold on to some of the developments he creates and sell others. Trading profits from the sales should help to cover the income shortfall on the properties that are retained (the difference between the interest on the money borrowed to develop them and the rents they produce) thus allowing the company gradually to build up an asset base.

The stock market's approach towards trading companies or those with a mix of trading and rental income is not always logical or consistent. Sentiment towards property trading companies can fluctuate very widely and so can their share prices, as was demonstrated by the stock market crash of October 1987. Shares as a whole fell well over 30% in this panic, but the shares of property trading companies generally fell considerably more.

Additionally, it is often true that when a property trading company begins to accumulate some assets in addition to its trading income, it does not get full credit for the added security in terms of its share price. The stock market seems to find it particularly difficult to evaluate a company which falls between the two stools of an asset-based stock and an earnings-based stock.

However, generating trading profits is not the only way of dealing with the income shortfall between rental income and the costs of borrowing. A whole range of other techniques has evolved to address the same problem, and we will be looking at them when we come to examine financing techniques in detail. Broadly, they usually require the property owner or developer to give up a proportion of his development profit or the future growth in income and asset values in return for finance at rates below those for a straight loan.

But here again there are choices to be made. A property developer can give away part of the profit from a single development or he can give away part of the future growth of the company as a whole by issuing shares or a deferred share stake in return for finance.

In general, investment companies are reluctant to finance new development by the issue of shares. First, the shares will normally need to be issued at a discount to asset value, thus "diluting" the assets (we will look at this in detail later). Second, future growth in the income from all the company's properties — and future increases in the value of these properties — will need to be spread over a larger number of shares, thus diluting the impact on each individual share. A trading company does not usually have quite the same worries about issuing additional shares, partly because its shares probably stand at a premium to net asset value.

Debt finance — borrowed money — can also be raised on the strength of a single development (project finance) or as a debt of the company as a corporate entity (corporate finance). And if it is raised as corporate debt by a stock-market-listed company, the borrower can chose between borrowings secured on the assets of the company and unsecured borrowings. But in practice only the larger and more established companies usually have the choice. The newcomer is not sufficiently well known as a corporate entity and is therefore probably restricted to raising money on individual projects.

Finally, there is the choice between fixed-interest borrowing and variable- or floating-rate borrowing (where the interest paid varies with changes in the general level of interest rates). An established investment company with virtually assured growth in income from its existing properties may be particularly attracted by the certainty that long-term fixed-interest borrowing brings.

15
Property and property company shares

Having looked at some of the factors affecting property financing, we need to see how they apply in practice in a property company structure. First, however, it is necessary to appreciate some of the fundamental differences between investing in property direct and investing in property company shares. Among these differences are:

- Rental income from a directly-owned building comes through to the owner without any deduction for tax, though he will need to pay tax later if he is liable. Income from a property owned by a company, after outgoings have been deducted, is liable to corporation tax (for illustrative purposes we are sticking with the 35% corporation tax rate that applied in 1990-91). Dividends are paid by the company out of income that has borne corporation tax, though the corporation tax charge covers all or part of the basic rate income tax that shareholders would otherwise have to pay on the dividend (it depends on the individual shareholder's tax status).

In other words, the company pays income tax on dividends (known as Advance Corporation Tax or ACT) on behalf of its shareholders, then offsets the cost against its own UK corporation tax charge. A shareholder who is not liable to income tax, like a pension fund, can claim back the income tax on the dividend but cannot claim back the mainstream corporation tax. To illustrate the principle, we first show the position as it would have been in 1990. Subsequent changes to the corporation tax and tax-credit system have introduced a complication which we will deal with in a moment. So, with

STOP PRESS— See "Company taxation" p352

corporation tax at 35% and basic rate income tax at 25%, the sums would have gone like this for a property producing rents of £1m a year, and assuming no expenses within the company:

Property owned direct

	£
RENTAL INCOME	1,000,000
RECEIVED BY OWNER	1,000,000

Property owned by company

	£
RENTAL INCOME	1,000,000
Less: CORPORATION TAX	350,000
NET INCOME AFTER TAX	650,000
NET DIVIDEND PAID	650,000
MAXIMUM INCOME TAX RECLAIMABLE	217,000
MAXIMUM RECEIVED BY INVESTORS	867,000

Thus, even for a shareholder who could claim back the whole of the income tax deemed to have been paid on the dividend, maximum income would only have been £867,000 against the £1m from owning the property direct. This is one reason why it may not suit non-taxpayers like pension funds to hold properties through joint-venture companies with developers and why it may be difficult (or tax-inefficient) to split the ownership of an individual property via a company structure.

However, since the original edition of "Property and Money" went to press in 1990, the government has introduced an added complication. Its effect is to penalise still further the ownership of property via a company for some classes of investor. Since the structure that applied in 1996 seems unlikely to represent the government's ultimate tax plans, we will simply describe what had happened up to that point.

The government decided that non-tax-paying shareholders could no longer claim back tax on dividends at the basic rate (25% in our examples). Instead, they would be able to claim back only at the new lower rate of 20%, which also became the rate at which companies paid ACT. This applied to dividends on all classes of shares and threatened to depress the income of the pension funds in particular. Moreover, it was

STOP PRESS— See "Company taxation" p352

unclear whether the reduction in the tax credit to 20% was the last move in this direction, whether it was merely a prelude to phasing out the tax credit system entirely, or whether it foreshadowed an eventual reduction to 20% in the basic rate of income tax.

Be that as it may, the situation in 1996 would have been as follows. The actual corporation tax rate in that year was 33% so the after-tax income of our company would have been £670,000 and we again assume that the whole of this was paid out in dividends. At a rate of only 20% for the tax credit, the maximum tax reclaimable fell to £167,500. The maximum income receivable by a non-tax-paying investor was thus £837,500 — rather worse than in our original example.

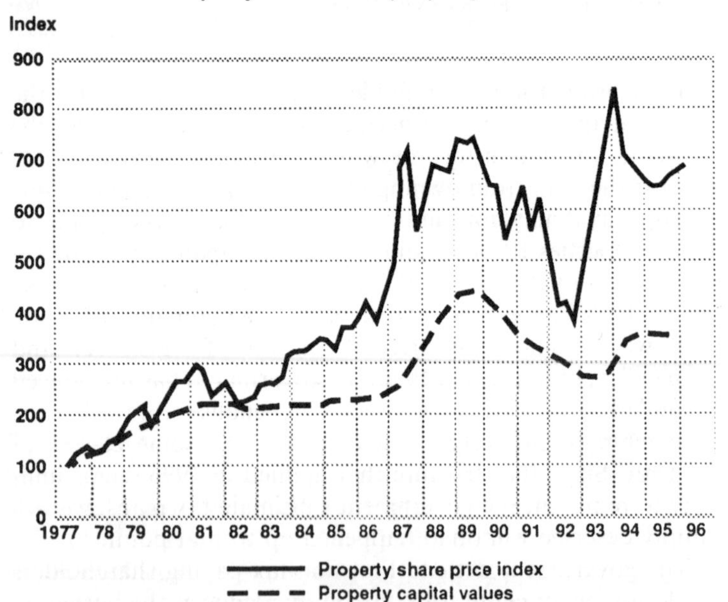

The picture you get depends on your starting point. But, since 1977 at least, property company shares, as measured by the Datastream property share index, have shown greater value growth than commercial property itself, as measured by Hillier Parker, despite some major upsets along the way. Giving the gearing built into most property companies, that is very much as it should be. Source: *Hillier Parker; Datastream*

STOP PRESS— See "Company taxation" p352

• The company structure poses still greater problems when it comes to capital gains. If a company sells an investment property at a profit it will be liable to tax on capital gains at the corporation tax rate. In practice the original cost of the property can be indexed to the cost of living to reduce the profit subject to tax, and capital gains which accrued prior to 1982 are tax-free. But for the purposes of illustration we will assume all the gain arose after 1982 and there was nil inflation. An individual is assessed to capital gains tax (CGT) at his maximum income tax rate (we will assume 40%). A pension fund normally pays no capital gains tax. Again, we look first at the position as it would have been in 1990

Property owned direct by pension fund

	£
COST OF PROPERTY	1,000,000
SALE PROCEEDS	2,000,000
SURPLUS	1,000,000
CAPITAL GAINS TAX	NIL
NET SURPLUS	1,000,000

Property owned by pension fund via a company

	£
COST OF PROPERTY	1,000,000
SALE PROCEEDS	2,000,000
SURPLUS	1,000,000
TAX AT 35% IN COMPANY	350,000
NET SURPLUS	650,000
MAXIMUM INCOME TAX RECLAIMABLE	217,000
MAXIMUM RECEIVED BY INVESTORS	867,000

The problem for the company is that it is taxed on realised gains on the sale of its properties, but the shareholder is also taxed on any profit on the sale of his shares if he is liable. The only consolation is that the company, if it distributes the surplus on property sales to shareholders, can offset the ACT on the distribution against the company's tax charge. Thus, in our example, if the whole net £650,000 profit had been distributed and if all shareholders had been non-tax-payers, these shareholders could have claimed back the £217,000 of ACT and their gross receipt would have been

STOP PRESS— See "Company taxation" p352

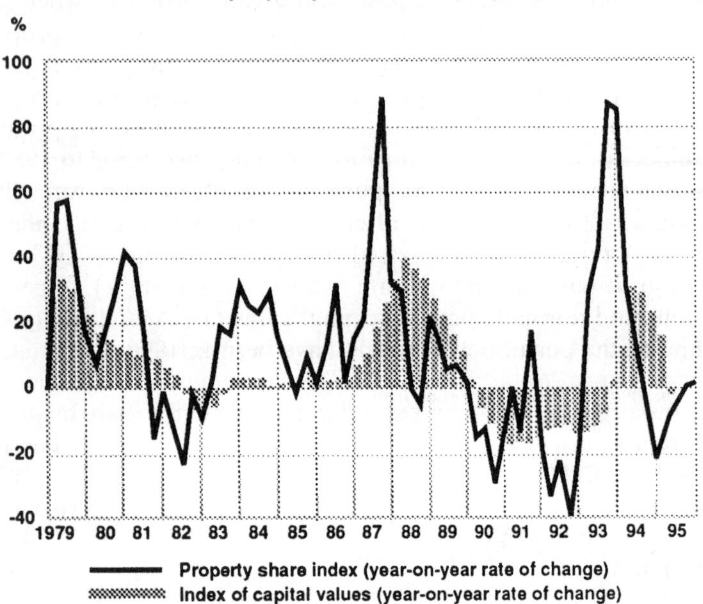

There was a time when the property world trumpeted the stability of commercial property values relative to other forms of investment. The slump of the early 1990s undermined that idea. But our graph shows that movements in the values of average investment properties are still generally far less extreme – and rapid – than movements in the value of property company shares, shown by the solid line. Source: *Hillier Parker; Datastream*

£867,000 — but still less than the £1m they would have got if they had held the property direct. If we again adjust for the tax changes between 1990 and 1996, the maximum gross receipts assuming that all shareholders are non-tax-payers would have come down to £837,500.

Now take the case where the company does not sell the property it owns but the shareholders sell their shares in the company instead. Suppose that the shareholders, instead of being pension funds, had been tax-paying individuals (paying at the higher 40% rate), that the company had 1m shares in issue, and that the value of the shares had risen

STOP PRESS— See "Company taxation" p352

from 80p to 160p to reflect the rise in the value of the property. If they had all sold their shares, additional tax of £320,000 would have been payable by the shareholders (1m shares at 160p = £1.6m; total profit on sale of shares = £800,000; CGT at 40% = £320,000). We are ignoring throughout these examples the possible reduction in the CGT charge from uprating the base cost of the shares or properties for inflation.

There are other important differences to take into account between direct and indirect ownership of property:

- The direct owner of a property has to undertake or arrange management of his investment. The shareholder in a property company has the management undertaken for him by the company.

- The stock market value of the shares of a listed property investment company is normally below the value of the assets attributable to each share (NAV or net asset value per share). Thus if a company owns properties of £10m, has 5m shares in issue and no borrowings, the NAV would be 200p. Typically, the shares might trade in the market at 160p or so.

  The reasons for this are the tax disadvantage from which the company suffers and the fact that, if the company were to dispose of its properties, capital gains tax might be payable and the net receipts could be below 200p per share.

  If the NAV is 200p and the shares trade at 160p, we would say they are at a "20% discount to assets". The discount varies between companies and changes with differing stock market conditions. At times when the stock market is booming and property values are expected to rise fast, the discount may temporarily disappear.

- Property company share values normally fluctuate far more widely than the values of properties themselves. The value of the shares will reflect to some extent the value of the properties the company owns, but it will not mirror it exactly. Property shares can be affected by stock market booms and slumps, regardless of what is happening in the property world itself (see chart on p. 284 for a picture of what happened to

STOP PRESS— See "Company taxation" p352

property company shares in October 1987, when property rents and values were actually rising).

• The stock market tends to anticipate events and property company shares will tend to anticipate events in the property world. Property company shares will often rise when property values are expected to increase. By the time the increases are being reflected in property company revaluations, the shares may have reached their peak and might even be moving down. Look how property company shares rose in the first half of 1987, before it was generally appreciated how fast rents and values were rising.

Because of this characteristic of property shares, they are often a useful guide to what may be about to happen in the world of property itself.

STOP PRESS— See "Company taxation" p352

16 Foreword to the figurework

There is an old boardroom joke which goes something like this. The directors of a company are hotly debating the merits of an expansion plan for the business. The company's chief accountant tries to contribute, but is shut up by the chairman: "Pipe down. You're just the scorekeeper".

Few companies would take that attitude today. The equivalent of the old chief accountant is almost certainly known as "finance director" and is a key figure in boardroom decision-making. He probably has a chief accountant under him.

But if the finance director plans financing policy, he also has to keep an eye on the score and think what impression it is going to make on shareholders and the financial community at large. The company's annual accounts are effectively the scoreboard where the company records what it owns and owes, and the profits or losses it has chalked up over the year. And finance directors have had to adapt to a number of changes in the method of presenting the score in the early 1990s.

This is because the accountancy world itself has changed. A new standard-setting body, the Accounting Standards Board, was established early in the decade with considerably greater powers than its predecessor. Two related bodies, the Urgent Issues Task Force and the Financial Reporting Review Panel, were established to help in interpreting the score-keeping rules and to clamp down on abuses.

The Accounting Standards Board took the view — rightly — that score-keeping practices had grown very lax in the 1980s. Companies had been using all manner of accounting tricks to present the score in a more favourable light than was justified. The accounts of some companies, while observing the letter of the accounting rules, diverged so markedly from the spirit that they seemed designed to mislead rather than inform the investing public.

Auditors — the independent accountants who are reqired to review company accounts and certify that they present a "true and fair view" (or not, as the case may be) — had also grown slack in many instances. Some severely misleading sets of accounts had been given a clean bill of health.

The Accounting Standards Board and its sister bodies set about tackling these abuses. In its first few years the Accounting Standards Board introduced a number of new standards, some of which had significant implications for score-keeping among property companies.

As this revised edition of "Property and Money" was being prepared for publication in 1996, some of these new standards had only recently come into force or were in the process of coming into force. It was thus too early to judge their full effect or to know whether they would need to be amended in the light of experience. Our coverage of property company accounting in this revised edition therefore reflects this transitional period.

Some of the more deceptive accounting practices we refer to in the chapters that follow will be more difficult if not impossible under the new accounting regime. Notably, it will be (and already is) considerably more difficult than in the past for a property company to disguise the full extent of its debts by using various tricks to keep them out of the accounts ("off balance sheet").

But accountants are an inventive lot, and doubtless they will try to find ways round the new rules as they did with the old ones. For this reason we have not completely re-written the accounting chapters that follow to eliminate references to pre-1990 practices that would be considerably more difficult today. We have, however, indicated where a particular practice might be outlawed under the new rules. We have included a new chapter (Chapter 25) on the cash flow statements that companies are now required to provide. And in the remainder of this chapter we set out the main accounting changes since 1990 that will affect the shape of accounts to come. If they seem a little complex and technical at first reading, they should fall into place once the following, more detailed, chapters on property company accounting have been absorbed.

We will start — as the Accounting Standards Board did — with cash flow statements. These have been a great bonus for users of property company accounts. We have already seen in Chapter 14 that "profit" can be a somewhat arbitrary concept, particularly among property companies. There is far less room for argument about cash. Companies may be able to exist for a time without making a profit. But if they run out of cash they are likely to go bust.

The difference between profit and cash is often misunderstood, so let us take an example. Trader Plc is a development company that sells its completed developments for a trading profit. Some of these developments are undertaken by associated companies: companies in which Trader has a significant stake but which it does not control. Let us suppose that one of these associate companies is called Splurge Developments and that Trader owns a third of its shares. Two other partners in Splurge Developments also own a third of the shares each.

Splurge Developments is carrying out a £150m town centre redevelopment. Trader has lent it £20m of the money that it needs at 10% interest. So Splurge is due to pay £2m of interest each year to Trader. Trader includes this £2m interest as part of its own profit under the heading of "interest receivable".

But note that word "receivable". In practice, Splurge has no spare cash until its development is completed and let. So it does not pay the £2m-a-year interest to Trader in cash. It simply adds it to the amount of money that it owes to Trader and will repay at the end of the day. Trader has, quite legitimately, treated the £2m as part of its profit, because the £2m has been earned during the year. But since it has not been received in cash, it would not feature in Trader's cash flow statement.

The requirement for companies to publish a cash flow statement (the precise form of which was amended by a revised standard in 1996) means that it is now far easier to see whether property companies are generating cash as well as profit. We stress again, failure to generate sufficient cash is why companies go belly-up.

The Accounting Standards Board's next important move was to amend the way in which companies report their financial performance with a new standard, Financial Reporting Standard 3 or FRS 3. In practice, this relates mainly to the shape of the profit and loss account.

Until FRS 3 came into force, companies had enjoyed considerable latitude in the way in which they treated certain items in their profit and loss account. Again, take an example. Trader Plc makes a profit of £10m before tax this year from its normal activities of developing commercial properties and houses for sale. But the housebuilding company, which it bought some years back, had been making hardly any profit and its position was getting worse. Therefore Trader decided during the year to close it down. This involved writing off the money Trader had originally paid for it, plus making large redundancy payments to the staff who have lost their jobs. The total of these items is, say, £8m.

Before FRS 3 came into force, Trader would probably have argued like this. "The £8m cost of closing down the housebuilding company is not a normal part of the costs of our business. It is a one-off item. There is therefore no reason why it should be allowed to affect the profits we publish for the year".

So, instead of knocking the £8m closure costs off its profits of £10m from other sources and reducing its published profits to £2m, Trader treated the £8m closure costs as an "extraordinary item". This meant that it published pre-tax profits of £10m and quietly wrote off the £8m from its profits after tax.

The problem with this treatment of "extraordinary items" was that companies abused it. The pleasant one-off items like profits on sale of subsidiary companies tended to be included in the published pre-tax profits as "exceptional items". The nasty items, like closure costs, were treated as "extraordinary items", prevented from affecting published pre-tax profits, and simply deducted from profits after tax.

Under FRS 3, virtually all one-off items like Trader's costs in closing its housebuilding business or profits on the sale of a subsidiary company have to be taken into account in calculating the pre-tax profit and the earnings of the company (see Chapter 21). Thus, under FRS 3, Trader's published pre-tax profits would be £2m, not £10m. FRS 3 also insists that published profits are broken down between profits from continuing businesses, profits from new companies acquired during the year and profits from businesses that have been discontinued or disposed of (like Trader's housebuilding arm).

This new treatment certainly prevents companies from hiding their costly failures. On the other hand, it can also make it more difficult to see what profits a company is capable of

earning in the long run. Published profits may fluctuate far more widely from year to year than they did in the past. Arguably, that £10m profits figure is a better guide to the profits Trader might be capable of earning next year than the £2m it would have to publish under FRS 3. Whether the FRS 3 standard will be amended in the future remains to be seen. In the meantime investment analysts tend to concentrate on the profits a company earns before taking account of the one-off items, whether favourable or unfavourable. Where they base their calculations of the company's earnings on profits that exclude one-off items, you may see these earnings referred to as "headline earnings".

FRS 3 also introduced a new financial statement — a "Statement of total recognised gains and losses" — to be published in addition to the profit and loss account. The profits earned by the company during the year would form part of these "gains". But the statement would also include surpluses or deficits thrown up by property revaluations during the year that were incorporated in the accounts. This new statement is very useful for users of property company accounts, since it highlights the overall result of the company's operations during the year and the amount of value created or lost for shareholders — not just realised profits but gains or losses from the change in value of the properties owned as well.

After FRS 3 came — not surprisingly — FRS 4. In this new standard the Accounting Standards Board sought to regulate the way that companies treated financial instruments (shares, loans, etc) in their accounts. One of the main effects of the standard was that it became far more difficult for a company to dress up its debts as anything else.

FRS 4 also saw to it that the accounts reflected more accurately than in the past the true cost of finance to the company. Suppose a company issues a loan that pays interest of 6% for the first three years, 12% for the next four years and 18% for the remaining three years of its ten-year life (a "stepped-coupon bond"). Under the old rules the company would have deducted interest at only 6% for the first three years. Under FRS 4 it has to calculate what compound rate of interest it is paying over the whole life of the loan, then charge interest at this rate each year. It means that the interest it has to account for is considerably higher than the 6% it actually pays in the first three years, though considerably lower than the 18% it has to pay in the last three years.

But it is the following standard, FRS 5, that probably has the greatest implications for property companies. FRS 5 is concerned with "the substance of transactions". In other words it says that, in deciding on accounting treatment, you need to look at what lies behind a transaction and not merely at its legal form.

Take an example. Trader Plc gets its merchant bank to set up a shell company with a share capital of just £100, owned by the merchant bank or one of its associates. The shell company does not therefore belong to Trader.

Trader then sells a property to the shell company for £50m. The shell company raises the £50m to pay for the property by issuing £50m of ten-year bonds to investors. The interest on the bonds is paid by the rents from the property.

The property had stood at £35m in Trader's own accounts, so Trader is able to chalk up a £15m profit on the deal. And at the same time as selling the property to the shell company, Trader arranged for the shell company to give it an option to buy the property back for £50m after ten years. In its turn, the shell company had an option enabling it to require Trader to buy the property back for £50m after ten years.

What is the substance of this transaction (and there were a number on these lines in the late 1980s)? If we look simply at the legal form of the transaction, Trader has sold a property to an independent company in which it has no shares. It has therefore disposed of the property and earned itself a £15m profit. And this is how the transaction would probably have been presented in the past. The £50m borrowed by the shell company via its bond issue would not, needless to say, have appeared as a debt in Trader's accounts. It was not money borrowed by Trader.

Now let us look at what really happened. In effect, Trader simply "parked" the property in the shell company for ten years. If the value of the property is over £50m in ten years, Trader will exercise its right to buy it back for £50m. If the property is worth less than £50m, the shell company will exercise its option to force Trader to buy it back for £50m. One way or the other, the property will land back with Trader at the same price as it sold it for and Trader will have the benefit of any rise in the value of the property over the ten years — or suffer from any fall in value. In other words, it enjoys the risks and rewards of ownership of the property, much as if it had never sold it.

And under FRS 5, Trader would not be deemed to have sold the property to an independent buyer. There would be no £15m profit on sale. And the £50m loan raised by the shell company against the security of the property would count as a borrowing of Trader itself and need to be shown in Trader's group accounts. However the transaction has been structured, the facts are that Trader has retained the risks and rewards of ownership of the property and has raised £50m of finance against the property. So that is what the accounts would be required to show.

The principles of FRS 5 cover many situations of this kind. For example, Trader might have decided to undertake developments through companies that it did not technically own or control. Its main purpose would have been to prevent the borrowings of these companies from appearing in its own accounts. But if the reality was that these companies took their orders from Trader in one way or another, and Trader enjoyed the benefit of their activities, then these companies would be treated as "quasi-subsidiaries" regardless of their strict legal status. Their assets and debts would need to be included in Trader's group accounts just as if they legally belonged to Trader. And a note to the accounts would give the details of these "quasi-subsidiaries".

Options of one kind or another are frequently used in property transactions — our example of the shell company was just one instance. FRS 5 says that you have to look at the nature of the options. If the commercial reality is that an option will almost certainly be exercised, your accounting treatment has to assume that it will be exercised. If the commercial reality is that it will almost certainly not be exercised, you account on the basis that it will not be exercised.

Take another example. A property company negotiates a "sale and leaseback" of a property with the finance arm of a bank. Thus, it "sells" the property but continues to have the use of it under a lease from the "buyer". It pays a "rent" geared in one way or another to money-market borrowing costs and has the option to buy the property back after five years at the same price as it sold it for.

Under FRS 5, this would not be treated as a genuine sale. The option to buy the property back at the original price will almost certainly be exercised. The reality of the transaction, however it is dressed up, is that the property company has raised a secured loan against the property. And this is what it will need to show. The property remains in its accounts and the money it raised against it is included in the company's debts.

Thus, FRS 5 looks as if it will outlaw many if not most of the tricks that property companies have used in the past to prevent the full extent of debts from appearing in the group accounts. The exceptions are when money is borrowed by a genuine joint-venture company that is not owned or controlled by the company in question.

The change is very important. Many of the property trading companies that collapsed with excessive borrowings in the 1990-93 period would never have been able to run up such large debts if the accounts had given a better picture of the true extent of their liabilities.

Two other aspects of accounting that could affect property companies were under review in the mid-1990s, though firm decisions were still some way off.

One concerns the use of valuations in company accounts. The Accounting Standards Board has suggested that certain classes of asset — including non-specialist property — should be regularly revalued. We have seen that properties held as fixed assets by property companies are already revalued on a regular basis. The new suggestion — if it comes into force — could require properties held under current assets (ie, trading stock) to be revalued as well.

Meantime, a discussion paper issued in 1996 suggested that industrial and commercial companies that wished to revalue the properties they used in their businesses should be required to do so on a regular, annual basis. They should also be required to provide depreciation on these properties. But property companies themselves were relieved at the recommendation that depreciation should not be required on investment properties.

The other area of possible future change concerns accounting for leases. We do not need to go into the detail. But under the standard that still applied in the mid-1990s, a distinction is made between "finance" leases and "operating" leases. The accounting treatment of the two is totally different. If a company uses assets held under a finance lease, in effect it has to account for the assets much as if it owned them direct. And the liability to make payments under the lease is treated much as if it were a straightforward debt of the company. This fairly rigorous treatment does not apply to operating leases.

Property companies have been able to exploit the distinction. Leaseback deals that were really financing transactions have been dressed up as operating leases rather than finance leases to prevent the money raised from showing as a debt in the group accounts.

As we have seen, this kind of trick might anyway be caught in the future under the provisions of FRS 5, which looks at the substance of a transaction. But the Accounting Standards Board might also decide to abolish the different treatment of finance and operating leases at some point in the future. A discussion paper from a working party of members of international accounting bodies has recommended precisely this course.

In this chapter we have inevitably run ahead of ourselves in touching on some fairly complex accounting and financing considerations. Their importance should become clearer after you have read the following chapters on property company accounts. But you might still find it useful to glance back at this chapter after covering the more routine aspects of property company "scorekeeping".

17
Tackling property company accounts

Evaluating property companies is a specialist art, and it is possible here only to give an outline of the main points to look out for. We have taken as our example (see page 106) an investment and development company with a smallish additional income from property trading. The accounts are simplified to emphasise the main features, and with most real-life listed property companies the figures would clearly be many times larger.

The balance sheet, which we will deal with first, is simply a snapshot of everything the company owns and everything it owes on one particular date — the last day of the company's financial year.

The investment properties of £10m that the company owns are shown under the heading of "fixed assets": the company intends to hold on to them, and they are not being bought or sold in the course of everyday business. It does, however, also own properties of £2m which are treated as trading stock and therefore included under the heading of "current assets". The investment properties under fixed assets will normally be shown at a recent valuation, whereas the trading properties are probably included at the lower of cost or realisable value. Cost will include expenditure on the properties since purchase plus, usually, interest on the money used to buy or develop them. They may be worth more than the figures shown in the accounts.

In addition to its £10m of investment properties the company has other fixed assets of £1m, giving a total under this heading of £11m. If we add in the current assets, including debtors (money owing to the company) of £600,000 and cash temporarily held in the bank of £400,000, the total assets of the company come to £14m. This is the sum of everything the company owns.

Consolidated Balance Sheet

FIXED ASSETS		£
Properties		10,000,000
Furniture and vehicles		1,000,000
		11,000,000
CURRENT ASSETS		
Stocks — trading properties	2,000,000	
Debtors	600,000	
Cash	400,000	
	3,000,000	
Less CREDITORS		
(amounts due within a year)	2,000,000	
NET CURRENT ASSETS		1,000,000
TOTAL ASSETS LESS CURRENT LIABILITIES		12,000,000
Less CREDITORS		
(amounts due after more than a year)		3,000,000
		9,000,000
CAPITAL AND RESERVES		
Issued ordinary share capital (£1 shares)		3,000,000
Revenue reserves		2,000,000
Revaluation reserves		4,000,000
SHAREHOLDERS' FUNDS		9,000,000
Net assets per share		300p

However, in the accounts the "creditors due within one year" (these used to be called "current liabilities") are usually knocked off the current assets to give a figure for "net current assets": in this case, £1m. In other words, short term liabilities are offset against short-term assets. The current liabilities include bank overdrafts or other short term money the company has borrowed. We need to go to the notes to the accounts to see the detail.

Assume in this case that the company has borrowed £1.5m of short-term money. The remaining £500,000 comprises: "trade creditors" (bills the company has not yet settled); the cost of the dividend it has proposed paying but which has not yet been paid; and corporation tax for which the company is liable but which is not yet due.

Adding together the £11m of fixed assets and the £1m of net current assets gives a figure of £12m for the company's total assets less its short-term liabilities. The final deduction to make is the £3m figure for "creditors due after more than one year". We will assume that all of this is accounted for by the company's long-term borrowings. Again, the notes to the accounts will give the detail. Let us say that the company has £2m of long-term mortgage debenture stock raised at 6% in the days of lower interest rates and a £1m long-term loan at a floating rate of interest.

Knock off these longer-term borrowings and we are left with a figure of £9m. This is the value (for accounting purposes) of everything the company owns once everything it owes has been deducted. This £9m that remains is the company's net asset figure and it is the book value of the shareholders' interest in the company.

Below this we see how the shareholders' interest in the company breaks down. There is £3m of ordinary share capital in issue: 3m £1 ordinary shares. There are revenue reserves (perhaps called "profit and loss account reserves") of £2m, representing profits earned in past years which were not distributed to shareholders but ploughed back into the business; they belong to shareholders all the same. And there is £4m of revaluation reserves, representing surpluses that have arisen when properties were revalued. The total of all these items is the shareholders' funds.

Finally, there is a "net asset value per share" or "NAV" figure of 300p. This is the book value of the assets attributable to each share in issue. The net assets amount to £9m, there are 3m shares in issue, so the NAV per share is £9m divided by 3m, or 300p.

What can we deduce from this balance sheet? The first item most analysts would look for is the gearing of the company.

This is the relationship between the borrowed money and the shareholders' money used in the business. We will come to its significance in a moment. In this case the company has £1.5m of short-term borrowings under the heading of creditors due within a year and £3m of long-term borrowings under the heading of creditors due in more than a year: a total of £4.5m of borrowed money.

Since there is £9m of shareholders' money in the business, the borrowings are equivalent to 50% of shareholders' funds. This is the most common measure of gearing. Sometimes net borrowings will be taken in place of gross borrowings, in which case we knock the £400,000 of cash off from the £4.5m of borrowings to give net borrowings of £4.1m. Alternatively, borrowings can be expressed as a ratio to gross assets.

The next point to look at is the composition of the group's borrowings. The £1.5m of short-term borrowings is almost certainly at variable rates of interest. If rates go up in the economy in general, the company will find itself paying more interest on these borrowings, which will tend to reduce its profits. Interest on the £1m long-term loan is also at a

floating rate of interest and the cost will go up as interest rates rise.

So of the £4.5m total borrowings, £2.5m will be affected by rises or falls in interest rates, unless the company has "capped" or otherwise hedged its borrowing costs — see Chapter 31. However, the remaining £2m of 6% debenture stock is a fixed-rate borrowing. It is very cheap money by recent standards, and the cost will not change as interest rates rise or fall elsewhere. A company with a lot of borrowings at floating rates of interest may see its share price suffer at a time when interest rates are rising.

Next we look again at that £10m figure for investment properties. When were the properties last valued? The notes to the accounts should tell us. Property companies are meant to value their investment properties each year and have a full independent valuation at least every five years, but it might still be almost a year since the last valuation, which could be significant in a period of rapidly rising or falling property values.

Moreover, if the company is developing properties to retain as investments, the £10m figure might include properties in course of development which are not yet completed and let. Such properties are likely to be shown at what they cost until they are completed and revalued, at which point it must be hoped that they will be worth significantly more than cost. The surplus over cost is value the company has created for its shareholders.

In addition, by comparing the volume of properties shown at cost (which are probably developments in train) with the figure for properties shown at valuation, we can get an idea of how active the company is as a developer. One final point to look for is the breakdown of properties between freeholds, long leaseholds and short leaseholds. By itself it does not tell us a great deal because a leasehold might be a traditional leasehold where the company pays a fixed ground rent, or it might represent a development financed by leaseback where the ground rent the company has to pay will probably rise as the rack rents rise. There should be some clues to this in the profit and loss account which we will come to later.

The trading properties under current assets are likely to be shown at cost anyway, and might be worth significantly more than the book figures.

Suppose we arrive at a guesstimate that the investment properties are likely to be worth £12m rather than £10m. What does this mean for shareholders?

This is where the gearing comes in. Our guesstimate is that an up-to-date valuation would increase the value of investment properties by £2m, thus raising total assets from £14m to £16m: a rise of 14.3%. So we mentally add the £2m to the value of investment properties in the books and add the same amount to revaluation reserves. This takes revaluation reserves up from £4m to £6m and increases total shareholders' funds from £9m to £11m. Dividing by the £3m shares in issue we find that assets per share have risen to 366.7p: a 22.2% increase over the 300p book figure.

So, because the company is partly using borrowed money to finance its operations, the value of the assets attributable to shareholders will rise at a faster rate than the company's assets as a whole. This is the beneficial aspect of gearing. Needless to say, it also works in reverse and any reduction in the value of the company's assets will result in a correspondingly larger reduction in assets attributable to shareholders.

Is there anything else we should be looking for in the balance sheet? One useful pointer is any changes between the previous years' figures and the latest ones. If the figure for investment properties has increased, is it simply because properties have been revalued upwards or is it because further amounts have been spent on developing or buying properties over the year? The notes to the accounts will help us to work this out, and the chairman's statement and directors' report which normally accompany the published accounts will probably refer to any major changes over the year.

Has the company issued additional shares over the year? Has it arranged any further major borrowings? How big a programme of developments does it have? Where are its properties located and how do they break down between office, shop and industrial? Somewhere in the published report and accounts and probably in the chairman's statement or directors' report there will be information on most of these points.

There will also be a heading of "capital commitments" in the notes which will show how much future expenditure was authorised or contracted for at the date of the accounts. The figure is not necessarily comprehensive because there may be plans for large-scale developments which are not yet sufficiently advanced for expenditure to have been committed. But if there are big developments in train we will also want to know where the finance is coming from to cover the cost — the company had only £400,000 of ready cash at its year-end.

Will it be issuing further shares for cash? Has it arranged new borrowings and, if so, on what terms? Again, the chairman's statement may help. Look also at the heading "contingent liabilities" in the notes. This should tell you if the company has guaranteed the borrowings of associated companies, which could add to the liabilities it shows in its own balance sheet.

A large-scale development programme brings risk, but it also increases the possible rewards. If the company can develop properties which are worth more on completion than they cost, it is adding to the value of the shareholders' assets. But in the short run, if it borrows the money for the developments and pays a high interest rate, the interest might exceed the income from the new buildings even though there has been a gain in asset terms.

If it issues shares for cash to fund the developments, the initial cost will be lower — the price of the dividend on the new shares — but any future increases in income or asset values will need to be divided among the increased number of shares, thus possibly "diluting" the rate of growth in the earnings and assets attributable to each share. By and large, investment companies will avoid raising cash by issuing shares. They would rather keep the share capital small and increase shareholders' funds by creating new value through developments or by benefiting from the rise in value of their existing properties.

Starting from the published accounts we can make our own estimates of the future growth in the earnings and assets of the company and also attempt to assess the risk implicit in the operation. Sometimes the chairman will give an estimate of the amount by which earnings are expected to grow over the next few years as a result of increases in rents on existing buildings. And if details of developments are given we can also make our own assumptions as to the benefit they will bring the company.

On the face of it, the company we have picked is relatively low risk. Most of its properties are investment properties. Its gearing or use of borrowed money is about average (very high gearing can be dangerous because it implies large interest charges which have to be paid whether the company is doing well or badly).

But no two property companies are the same and each has to be separately assessed. We will be looking in the following chapters at some additional points that can crop up in the balance sheet and also at the company's income picture: the profit and loss account.

18
On and off balance sheet

One or two complications that crop up in property company balance sheets need to be dealt with before we move on to the profit and loss account.

The balance sheet we looked at in the last chapter was a "consolidated" balance sheet. Most companies of a size to have a stock market listing are, in fact, groups of individual companies, all ultimately controlled by the "parent" company or "holding" company. In a consolidated balance sheet, all of the companies constituting the group are lumped together and treated much as if they were a single company for accounting purposes.

Thus Payola Properties Plc might in fact consist of Payola Properties Plc, Whiz Estates Ltd and Dicey Developments Ltd. Payola is the parent company and controls the other two which are subsidiaries. If Payola itself owns properties of £5m, Whiz has properties of £3m and Dicey's portfolio is £2m, the figure for properties that will appear in the consolidated balance sheet of Payola Properties Plc is £10m. Likewise, the debts of the different constituents of the group will be aggregated.

In addition to subsidiary companies, Payola may also have "associated companies" or "associates". These are likely to be companies in which it has a share stake of over 20%, but below the 50.01% which would give control, and where it exerts some management influence. Suppose Payola had undertaken a development with two other parties in a joint company — call it Verruca Ventures Ltd — in which each had an equal one-third share.

Verruca would be an associate of Payola and in its own group accounts Payola would show its share of the net assets of Verruca. Suppose Verruca's accounts show properties of £7m, borrowings from banks of £4m and equity assets of £3m. In its own consolidated accounts, Payola simply shows under the heading of "associates" or "associated companies" its share of the net assets of Verruca. Verruca's net assets are £3m, Payola owns a third of the company, so it includes in its own accounts a figure of £1m for its interest in Verruca. The fact that Verruca has large borrowings is not reflected in Payola's accounts. Note, however, that the position could be changing as a result of proposals from the Accounting Standards Board (see Chapter 16) and that more information on the borrowings of associated companies may need to be given in future.

Finally, Payola might have subsidiary companies which it controls but where it does not own 100% of the share capital. Suppose that Payola owns 70% of the capital of Semolina Estates Ltd, but the founding Semolina family which sold control to Payola kept a 30% stake for itself. Payola will group Semolina Estates into its consolidated accounts, but account for the fact that other parties have a 30% interest with an item for "minority interests" or "outside shareholders' interests".

So much for the theory. In practice the definitions of subsidiary and associated companies have often been deliberately distorted, by property companies in particular. The reasons are as follows.

In the 1980s it was possible to borrow a high proportion of the cost of a development (arguably too high a proportion for what is still a fairly high-risk operation). It was also possible to borrow on "non-recourse" or "limited-recourse" terms. "Non recourse" means that the security for the lender is the individual development and whether or not he gets his money back therefore depends on the success of that development. He does not have a claim on the other assets of the group which is undertaking the development. "Limited recourse" usually means the parent does not guarantee the capital amount of the loans, but provides some guarantee of completion of the project and of interest payments.

If a development financed mainly by limited-recourse loans is undertaken by a subsidiary company, the group accounts will have to include the full extent of the borrowings the subsidiary has assumed. The group as a whole may therefore look very highly geared, in a way that could worry investors or bankers and limit its further borrowing power.

But if the development is undertaken by an associated company or by a company which is to all intents and purposes a subsidiary but manages to avoid falling within the technical definition of a subsidiary, the borrowings will not need to

appear in the balance sheet for the group. (However, note again that the latest accounting changes could greatly reduce the scope for disguising what is really a subsidiary company as something else.)

The group may thus look fairly low geared in its accounts, even though there are massive borrowings within the development companies. While in theory the parent company is not responsible for limited-recourse loans raised by other group companies, in practice it would not do a listed property company's image any good if it simply walked away from an unsuccessful development and left the lenders with the losses.

Raising money in such ways that it does not appear in your accounts is known as "off balance sheet" financing and became very prevalent in the late 1980s, though it will be more difficult under the new accounting rules. Off-balance-sheet financing means that published group accounts may give very little idea of the real scale of the group's assets and commitments. It has been most common with development and trading companies of the type often referred to as "merchant developers" (see Chapter 24).

The traditional definition of a subsidiary company was, very broadly, one where the parent has a shareholding and controls the composition of the board of directors or where it has more than 50% of the nominal equity capital. We do not need to worry about the detail of how companies got round this definition. Sufficient to say that a recognised class of companies grew up known as "controlled non-subsidiaries" which were effectively controlled by a parent but did not need to be consolidated into the group accounts.

To counter this trend a new Companies Act considerably extends the definition of a subsidiary, and means that the former controlled non-subsidiary now probably needs to be consolidated (included in the group accounts).

The accounting authorities, too, now require that accounts should reflect the commercial effect of a transaction rather than the mere form. Disguised subsidiaries will in future probably qualify as "quasi-subsidiaries" and need to be treated as if they were true subsidiaries in form as well as in substance (in fact, additional detail may need to be given on them). But developers are an inventive bunch, and will doubtless find loopholes in whatever new rules are applied. At all events, a company owned 50-50 with another party, where each party has the same number of directors and votes, will probably not count as a subsidiary of either.

The other common and older-established form of off balance sheet financing is the sale and leaseback. We examine this in more detail in Chapter 42. For the moment we only need the broad outline.

A developer sells the site for a development to an institution, which will put up the finance for construction. The institution grants a lease of the completed building to the developer at a rent of, say, 70% of the rack rental value of the building. The developer thus has a "top slice" profit rent of 30% of the total rents if it manages to let the whole building. The rent which the developer pays to the institution will probably rise in proportion as the rack rents rise.

The point about this arrangement is that, though it takes the form of a lease, it creates a liability in much the same way as if the developer had borrowed the money for the development. The developer is committed to paying the rent to the institution, whether the building is fully let or not. If the building should fall empty, this would be a very hefty liability.

But in a property company's balance sheet you are very unlikely to find an item reflecting this liability. The accountancy rules say that in the case of a finance lease (and a sale and leaseback might be regarded as a finance lease, though is usually treated as an operating lease) the balance sheet should show the leased item as an asset and the amounts due under the lease as a liability. In practice, if a property company has substantial liabilities under leaseback arrangements you are only likely to spot them from the size of the "rents payable" item offset against income in the profit and loss account. This indicates that the company is in practice very much more highly geared than you would deduce simply by totting up the borrowed money in the balance sheet.

But note again that the accountancy bodies are considering changes to the accounting rules for leases, and the distinction between finance leases and operating leases might disappear at some future point.

19
Classes of capital and capital issues

So far, when we have talked about the share capital of a company we have talked entirely in terms of ordinary share capital or "equity". These are the shares whose holders share fully in the risks and rewards from the operations of the business, and the dividend and the share price normally move to reflect the success or failure of the business.

But there are other classes of share capital that the company may have, and by far the most common of these is preference capital. For tax and other reasons it was out of favour for many years and most of the preference shares in company accounts were there for strictly historic reasons. But preference has been making a comeback in recent years.

Preference shares are share capital which has most of the characteristics of debt. They normally pay a fixed dividend, like a fixed rate of interest on a loan. The difference is that the preference dividend is paid out of income that has borne corporation tax, whereas loan interest is offsettable against profits for tax purposes. And a preference shareholder cannot put a receiver into a company to recover his money, whereas the holder of a loan can do so if the borrower defaults on the terms of the loan.

The precise characteristics of any particular issue of preference shares depend on the constitution of the individual company. But a typical traditional preference share might:

- Pay a fixed dividend on which (as with an ordinary share) basic rate income tax has been paid before it reaches the shareholder.

• Rank before the ordinary shares for dividend payment (the preference dividend must be paid before the company pays any dividend on the ordinary shares — so the preference dividend is far more secure than the ordinary dividend).

• Rank before the ordinary shares in a winding up. Preference shares are typically of £1 nominal value, and if the company has to be liquidated, preference shareholders would normally get their £1 plus arrears of dividend before ordinary shareholders get anything. So preference shares rank after the debt of the company but before the equity capital. In practice, in a winding up there may not be anything left for either the preference or the ordinary shareholders after the company's debts have been dealt with.

So, from an investor's point of view, preference shares are more secure than equity but less secure than a loan stock or a bond. And in their basic form they do not share in the increasing prosperity of the company.

However, many variations on the preference theme are encountered. You might come across "participating preference shares". These probably pay a fixed dividend but in addition pay an extra dividend related to the amount paid on the ordinary shares. And convertible preference shares are now common; we will be looking at these later.

But in a property company the most important point is how to allow for preference shares when making a net asset value (NAV) calculation. Take the following breakdown of shareholders' funds:

CAPITAL AND RESERVES	£
Issued ordinary share capital (£1 shares)	3,000,000
Issued preference share capital	1,000,000
Revenue reserves	2,000,000
Revaluation reserves	4,000,000
SHAREHOLDERS' FUNDS	10,000,000
Net assets per share	300p

Total shareholders' funds are £10m. But £1m of this represents preference capital, so ordinary shareholders' funds are only £9m, which is the relevant figure for NAV calculations. Since there are 3m £1 ordinary shares, net assets per share amount to 300p (£9m divided by 3m shares).

A couple of other technicalities relating to shareholders' funds need dealing with. First is the "scrip issue" or "capitalisation issue", which often causes confusion in its effects on assets per share.

Suppose we take a company with shareholders' funds identical to those above, but without the preference capital. There are 3m ordinary shares in issue and the NAV is 300p. And let us suppose the price of the shares in the stock market is 240p.

The directors decide that a share capital of £3m is rather small in relation to total shareholders' funds of £9m. So they decide to use part of the reserves of the company to create new shares, which will be issued free to shareholders *pro rata* with their existing holdings.

In this example the company decides to use £1m of its revenue reserves to create 1m new £1 ordinary shares. These are distributed free to shareholders in the ratio of one new ordinary share for every three they hold. So revenue reserves

go down from £2m to £1m and issued ordinary share capital goes up from £3m to £4m. Note that this is purely a book-keeping transaction. No new money is involved. A share-holder who had three shares before has four shares now. But since ordinary shareholders' funds are still only £9m, the NAV per share comes back to 225p.

After the scrip issue the breakdown will look like this:

CAPITAL AND RESERVES	£
Issued ordinary share capital (£1 shares)	4,000,000
Revenue reserves	1,000,000
Revaluation reserves	4,000,000
SHAREHOLDERS' FUNDS	9,000,000
Net assets per share	225p

Likewise, the share price adjusts for the scrip issue — assume it was 240p before the issue was announced:

Shareholder starts with:
3 shares at 240p = 720p
Ends with:
4 shares at 180p = 720p

So the market price adjusts down from 240p to 180p, but the total value of the shareholder's investment remains unchanged.

A rights issue, however, is a very different matter. This is when a company issues new shares for cash, offering them first to existing shareholders, and it increases the total share-holders' funds of the company. The shares are almost cer-tainly issued below the current market price to encourage shareholders to subscribe.

So let us go back to our example of a company with equity capital of £3m, ordinary shareholders' funds of £9m and a NAV of 300p. The market price of the shares is 240p and 1m new shares are offered at 200p in the ratio of one new share for three held. The issue thus raises £2m. For every new share issued at 200p, 100p represents the nominal value of the share and 100p is the premium at which the share is issued. In accounting terms nominal capital thus increases by £1m and the other £1m raised goes to a form of reserve known as a "share premium account". After the issue, shareholders' funds look like this:

CAPITAL AND RESERVES	£
Issued ordinary share capital (£1 shares)	4,000,000
Revenue reserves	2,000,000
Revaluation reserves	4,000,000
Share premium account	1,000,000
SHAREHOLDERS' FUNDS	11,000,000
Net assets per share	275p

The problem is that the issue of shares at a price below net asset value has diluted the NAV from 300p to 275p, and this is a very different matter from the technical adjustment in the NAV that occurs with a scrip issue. It is one reason why property investment companies are normally very reluctant to raise fresh capital via a rights issue.

After a rights issue, the market price of the shares adjusts to allow for the fact that new shares were issued below market value (the scrip element in the issue). In our example:

Shareholder starts with 3 shares at 240p	= 720p
Buys 1 new share at 200p	= 200p
Total for 4 shares	= 920p
Therefore value of 1 share	= 230p

Thus 230p is the theoretical "ex-rights" price. This is the price to which the shares should adjust down in the market on the day when buyers of the shares no longer acquire them with the right to subscribe for the new shares. But if the shares revert to a 20% discount to the new NAV of 275p, the price in the market would actually need to come back to 220p.

20 Calculating with convertibles

Our original sample balance sheet did not include one feature which has become increasingly common in company financing, and particularly property company financing: the convertible loan stock.

A convertible is mid-way between debt and equity finance. A typical convertible might be a loan stock carrying an interest rate — a coupon — of, say, 7.5% and with a life of 15 years. Initially it simply pays an interest rate of 7.5% (which the issuer can offset against profits for tax purposes) and appears in the accounts of the company as a liability like any other form of bond or debenture.

But between certain dates, holders have the option to convert the convertible loan stock into ordinary shares of the company on terms fixed at the time of issue. After all the stock has been converted (if it is), it disappears from the balance sheet as a liability and the share capital increases by the number of new shares created on conversion. If conversion does not prove worthwhile, after the end of the conversion period it is either repaid or reverts to being an ordinary loan stock and is redeemed at some later date.

Though traditionally convertibles were loan stocks, in recent years we have seen a number of convertible preference share issues. These work in much the same way: the convertible preference shares pay a fixed dividend until they are converted into ordinary shares. But since they are technically share capital rather than debt, the income takes the form of a dividend rather than an interest payment and has to be paid out of income that has borne corporation tax — shareholders get a credit for basic rate income tax paid, in the same way as with the dividend on an ordinary share.

Because the value of a convertible will at least partially reflect the performance of the ordinary shares of the company, investors are prepared to accept a lower rate of interest than they would require on a straight loan.

On the other hand, the yield will almost certainly be higher at the outset than the yield on the ordinary shares, and this can give the convertible defensive qualities at a time when the ordinary share price may be volatile.

Companies tend to look on convertible stocks as a cheap source of finance. Viewed as debt, the convertible costs less in interest than a straight loan stock. Viewed as deferred equity, it is seen as a way of issuing new shares at a premium to the current market price, which helps to get round the asset dilution problem that usually arises with a straight rights issue in ordinary shares. Typically, when a convertible is first issued the terms provide for it to convert into ordinary shares at 10% to 15% above the then current market price.

Suppose the company's share price is 175p. It might issue a convertible loan stock that converted into ordinary shares on the basis of one share for every £2 nominal value of loan stock. So the conversion price is 200p and if the company issues £2m nominal of the convertible loan this will ultimately (assuming full conversion) convert into 1m new ordinary shares.

In practice some investment analysts argue that companies are kidding themselves if they think convertibles provide cheap finance. Admittedly, there is the possibility of issuing shares at above the current market price. But the other side of this coin is that the company is paying interest on the convertible stock, which is considerably higher than the dividend on the ordinary shares, until conversion takes place.

If the company's share price rises fairly consistently in the years after the convertible is issued, a point is reached where the share price moves well ahead of the price at which the stock converts (and the value of the convertible stock in the market will, of course, rise to reflect this fact). Take our example of a company with a share price of 175p that issues a convertible that converts at 200p. Assume that five years later the share price has risen to 240p and stands at a discount of 20% to the company's net asset value of 300p.

An abbreviated balance sheet looks like this:

	£
TOTAL ASSETS LESS SHORT-TERM LIABILITIES	14,000,000
LONG-TERM SOURCES OF FINANCE	
6% Debenture stock	2,000,000
Term loan	1,000,000
5% Convertible loan	2,000,000
	5,000,000
CAPITAL AND RESERVES	
Issued ordinary share capital (£1 shares)	3,000,000
Revenue reserves	2,000,000
Revaluation reserves	4,000,000
SHAREHOLDERS' FUNDS	9,000,000
Net assets per share	300p

An investment analyst will tend to look at the effects of conversion on asset value, even though the convertible stock has not yet been converted. So he will draw up a balance sheet showing what the position would be after full conversion. It looks like this:

	£
TOTAL ASSETS LESS SHORT-TERM LIABILITIES	14,000,000
LONG-TERM SOURCES OF FINANCE	
6% Debenture stock	2,000,000
Term loan	1,000,000
	3,000,000
CAPITAL AND RESERVES	
Issued ordinary share capital (£1 shares)	4,000,000
Revenue reserves	2,000,000
Revaluation reserves	4,000,000
Share premium account	1,000,000
SHAREHOLDERS' FUNDS	11,000,000
Net assets per share	275p

The £2m of convertible has disappeared as a liability, boosting shareholders' funds by £2m to £11m. Of this increase, £1m is the nominal value of the new shares issued, taking issued capital up from £3m to £4m, and £1m is accounted for in the share premium account, since effectively shares of £1 nominal value were issued at £2. But though net assets are now up to £11m, they have to be divided among 4m shares, which gives a net asset value of 275p. So conversion dilutes net assets from 300p to 275p; this would be referred to as the "fully diluted NAV". Similar sums can be done to calculate the dilution effect on earnings (if any).

It is not uncommon nowadays for convertible stocks to be traded in the euromarkets rather than the domestic stock market ("euroconvertibles"), and they are often considerably more complex than the example illustrated here. In particular, they may incorporate an investor "premium put" option.

In essence, this normally guarantees the investor a minimum return close to that on government stocks. The convertible carries a coupon of, say 7.5%. But after a certain period investors have the right to sell the stock back to the company at a price which provides them with an overall return of, say, 10% a year compound from the date of issue. So if the share price rises far enough to make conversion worthwhile, the investors convert. If the share price performs badly they can exercise the put option instead, meaning they get a return close to what they would have had from a conventional non-convertible bond (see Chapter 36 for more detail).

21
The profit and loss account

Having dealt with the balance sheet we now turn to the other main component of a property company's annual report and accounts: the profit and loss account. This shows the revenue position over the year as a whole, and normally it will be a "consolidated" profit and loss account in which the results of all the companies in the group are aggregated as if these companies were a single entity.

Note that the profit and loss account shows the profit or loss according to the accepted accounting conventions. It does not necessarily show what the cash flows into and out of the company were — an important difference, as will emerge when we consider the treatment of interest charges. Investment analysts used to devote much effort to looking beyond the profit and loss figures to establish the company's cash flow. Their task is made considerably simpler now that companies themselves have to publish a cash flow statement (see Chapter 25).

The vital point with any property company is to see where its income comes from. We have already stressed that there is all the difference in the world between stable and increasing income from rents and the profits derived from property trading. "Manufacturing" property for sale is much like any other manufacturing operation and rather less dependable than most. Profits can fluctuate widely in any one year according to the buildings that happen to be completed and sold during the period.

Much press comment ignores this distinction and greets a one-off jump in property trading profits as if it necessarily indicated a rise in the company's long-term earning potential. It does not. Even trading companies with a record of good profits stretching over a number of years found that their income dried up completely in the property market crash of 1990-93. The investment companies deriving most of their income from rents survived, by and large, relatively unscathed.

Consolidated Profit and Loss Account

INCOME FROM INVESTMENT PROPERTIES	600,000
PROFIT FROM PROPERTY TRADING	100,000
	700,000
Less INTEREST PAYABLE	400,000
PROFIT ON ORDINARY ACTIVITIES BEFORE TAX	300,000
Less CORPORATION TAX	105,000
NET PROFIT FOR SHAREHOLDERS	195,000
Less DIVIDENDS	130,000
RETAINED PROFIT	65,000
Earnings per share	6.5p
Dividend per share (net)	4.33p
Dividend per share (gross)	5.41p

The income of our sample company (see simplified accounts shown above) is firmly based on rents, which account for £600,000 of the £700,000 income before interest. We know from the balance sheet that the company has £2m of properties held for trading, and the smallish £100,000 trading profit suggests that this has not been a very active year. Look back at the previous years' figures and see what average level of trading profit the company generates. It may indicate that trading profits are likely to be higher again in future years. But clearly investment is the company's main business and trading is a sideline.

You will probably need to go to the notes to the accounts to get more information on the make-up of the company's income from investment properties. What is the figure for gross rents? Is there a big deduction for management charges? Perhaps more important, do ground rents absorb a significant proportion of gross rents? If they do, it probably means that the company has made extensive use of leaseback finance in the past, and that part of its revenue may be "top slice" income. This is a form of income gearing which needs to be taken into account when assessing the total gearing of the company.

If ground rents are a small item, they probably represent rents paid on traditional leasehold properties. The details of properties in the notes to the balance sheet will have given a break-down between freeholds and leaseholds, but will probably not have indicated the type of leasehold, except to the extent that it is a short leasehold or a long leasehold. Income from freeholds or traditional long leaseholds is normally more prized by investors than top slice income arising from leasebacks.

The "interest payable" item also repays a little research. Does this represent the company's total interest bill for the year? Interest charges may be treated in different ways. The interest on money borrowed to undertake developments is frequently treated as a capital item and added to the balance sheet cost of the development properties until they are completed and revenue-producing. And it is still more common for the finance charges on trading properties to be treated in this way.

This can be very important for a company with a big development programme. If the £400,000 interest charge represents only interest on borrowings associated with the completed and revenue-producing properties, the company is in practice very much more highly geared in revenue terms than is immediately apparent from the profit and loss account.

Suppose the company in our example incurs a further £100,000 of interest which is added to the cost of developments in its investment properties ("capitalised", or treated as part of their capital cost). And suppose there is a further £200,000 of interest that represents the financing cost of the trading properties and is simply added to their book value. The company's real interest costs work out at £700,000, not £400,000, and would thus absorb the whole of its income for the year if charged to the profit and loss account.

Does this matter? In good times it does not. When the developments are completed, or soon after, the £100,000 of interest relating to them will have to be charged against the year's income. But by this time the developments should themselves be producing revenue to offset the interest cost. And provided the trading operations are successful, it will not matter that interest costs have been added to their book value. They will be sold at a figure that shows a profit even after allowing for this factor.

The problems arise if the new developments included in investment properties fail to find tenants on completion. After a time the auditors will insist that the interest charges relating to these buildings must be charged in the profit and loss account even though they are not producing revenue, and this will hurt the company's published profits. And if the property market hits a bad patch and values fall, the trading properties might turn out to be worth less than their book cost figure which has been inflated by capitalised interest charges each year. This was very clearly borne out in the 1990-93 market collapse when property trading companies were forced to write down their properties (reduce their value in the accounts) by massive amounts and frequently showed themselves to be insolvent.

In the early 1970s companies with very large development programmes frequently capitalised interest charges so that, via various accounting contortions, the revenue account still showed a profit. When the property market collapsed in 1974-76, they were forced to show the real interest position in the profit and loss account — ie, charge all the year's interest against revenue — and many companies emerged as heavily loss-making. A simular process took place in the market collapse of 1990-93.

So the treatment of interest is vital with any property company. The company in our example shows a pre-tax profit of £300,000, but if it charged all interest against revenue it would show no profit at all. There is no "right" way to treat interest on developments, but remember that if you are comparing one property company with another you may not be comparing like with like if they account for their interest charges in different ways. Companies which charge all financing costs against current revenue may be described by commentators as "conservative" in their accounting policies. The section on "accounting policies" in the annual report and accounts should make clear how interest is treated.

However, this possible source of confusion might disappear at some point in the future. In a discussion paper late in 1996, the Accounting Standards Board declared its intention of imposing a standard treatment. Either all companies would capitalise development interest or none of them would. But at that point the ASB had not yet decided which.

After corporation tax of £105,000 our company has net profits or "earnings" of £195,000, of which it pays out

£130,000 to shareholders by way of dividend. Divide the £195,000 of earnings by the 3m shares in issue and you find that the company earned 6.5p for each share; this is the "earnings per share" or "eps" figure. Undertake the same exercise for the dividend payment and you find the dividend per share comes out at 4.33p.

This is the net dividend per share figure — in other words, basic rate income tax (25% in our examples) has been paid and offset against the company's own corporation tax charge before the shareholder gets his dividend cheque. The company's corporation tax charge of £105,000 thus includes the basic rate income tax of £43,000 paid on behalf of the shareholders (this is known as advance corporation tax or ACT).

For purposes of comparison, most returns on money in the investment world are expressed gross (before tax) rather than net (after tax), as the actual amount any individual investor receives depends on his or her tax position. So we have to "gross up" the net dividend of 4.33p to find out the equivalent amount before tax.

This has become slightly more complex as a result of the tax changes we examined in Chapter 15. The position after these changes — and we would stress again that this might be only an interim phase — was that companies had to pay tax (advance corporation tax or ACT) to the Inland Revenue on profits that they distributed as dividends but only at a rate of 20%. Shareholders who were not liable to tax (like the pension funds) could, correspondingly, only claim back tax at 20% on these dividends. If they received a dividend of 100p, they could claim back 25p to give them a total of 125p. This is because 125p before tax is the figure which provides 100p net after deducting tax at 20%. We would therefore say that "125p is the gross equivalent of a net dividend of 100p". Confused? You are not alone.

In the same way, the gross equivalent of a net dividend of 4.33p is 5.41p because 5.41p is the figure before tax which will give you 4.33p after tax at 20%. How do you work this out? You could say that, since the tax credit is 20%, the net dividend is 80% of the gross figure you

STOP PRESS— See "Company taxation" p352

are looking for. So if you multiply the net dividend by 100 and divide by 80, you will arrive at the gross dividend. A simpler way of doing it is to divide the net dividend by 0.8 to get the gross figure.

However you do the calculation, the 5.41p gross dividend is the figure needed to calculate the yield on the shares. We will look at the calculation itself in a moment.

But before calculating the yield, we should examine the sources of the company's income. We have already seen that trading income is usually less reliable than income from rents, and one of the sums that the property company analyst will normally undertake is to see to what extent the dividend is covered by rental income. If the company did not make any trading profits in a particular year, would it still be earning enough revenue from rents to cover a dividend at the current rate?

In this case we take the £300,000 of pre-tax profit and knock off the £100,000 of trading profit to leave £200,000. Deduct corporation tax at 35% (£70,000) and we are left with £130,000. As it happens, this exactly covers the cost of the dividend so we are safe in saying that the company in no way relies on its income from trading to pay the dividend. This means that the dividend should be exceptionally secure. Companies which derive most of their income from trading normally cannot offer the same security and they may tend to pay out a smaller proportion of their earnings by way of dividend.

If you are analysing a property company you are interested in its likely future income as much as the present position. Look for any information the chairman gives on increases in income expected from rent reviews and reversions. And if you know there are major developments coming up to completion you can make your own estimates of the amount of income (net of financing costs) that they are likely to add to the revenue account.

If you know that the company is financed partly on variable-rate borrowings you may undertake calculations as to the likely effect of rises or falls in interest rates. Rising rates

STOP PRESS— See "Company taxation" p352

will increase the interest charge and tend to depress profits, temporarily at least.

Finally, you will want to see how the company ranks on the conventional stock market criteria. Remember that a property investment company is normally evaluated on an asset basis rather than a price earnings ratio (PE ratio) basis, though the price earnings ratio is the main yardstick for a company that derives most of its income from property trading.

Assume that the shares of our sample company are trading in the stock market around 240p. What dividend yield do they offer a purchaser?

We know that the gross equivalent of the dividend paid is 5.41p per share. To get the yield, simply express this as a percentage of the market price. The sum is:

$$\frac{\text{GROSS DIVIDEND}}{\text{MARKET PRICE}} \times 100 = \text{YIELD}$$

or in this case:

$$\frac{5.41\text{p}}{2.25\%} \times 100 = 240$$

We have already calculated from the balance sheet that the net asset value per share (based on the book figures) is 300p. This means that at a price of 240p the shares are trading at a discount of 20% to assets and yield 2.25% — fairly typical for the property sector in "normal" times.

STOP PRESS— See "Company taxation" p352

22
Profit and loss refinements

We looked in the previous chapter at a simplified property company profit and loss account. But, as with the balance sheet, there are some complications that can crop up in practice and we need to look at the more common ones. Many of them are counterparts of items we have already examined in the context of the balance sheet.

The first item you may come across which was not featured in our sample profit and loss account is "income from associated companies" or "income from related companies". Remember that an associated company or "associate" is not a subsidiary but usually a company in which the head company has a shareholding of between 20% and 50% and over which it exerts some management influence.

In the consolidated profit and loss account of the head company it takes in its proportionate share of the profits of the associated company or "associate". So if Payola Properties and two other companies set up Verruca Ventures to develop the Toytown Estate in London's docklands, and if each took a one-third share in Verruca, Payola would be entitled to one-third of Verruca's profits, and this is the figure it would show in its consolidated profit and loss account. So if Verruca made £90,000 pre-tax, Payola would include £30,000 for its share, and would show it as part of its income before tax.

But note that £30,000 (or its equivalent after tax) is not necessarily the cash that Payola receives from Verucca. Verucca may have needed to hang on to its profits to finance further development and may therefore have declared a dividend much smaller than its profits, or may have paid no dividend at all. Watch this point when a company derives much of its profit from associates.

Next, minority interests. Payola, remember, owns only 70% of its Semolina Estates subsidiary and the founding Semolina family hang on to the remaining 30%. They are, therefore, entitled to 30% of Semolina's profits. So in its consolidated profit and loss account Payola includes the whole of the profit of Semolina Estates. But before it can arrive at the earnings that belong to Payola shareholders it has to knock off 30% of the profits that Semolina contributed. It does this by making a deduction from its net profit after tax, equivalent to 30% of Semolina's after-tax profits. This is the profit after tax belonging to the 30% minority shareholders in Semolina.

A similar deduction has to be made if Payola Estates itself has preference share capital in issue. Suppose it has £500,000 of 8% (net) preference shares. The dividend on these will be paid out of profits that have borne corporation tax and will amount to £40,000. So this amount also has to be knocked off Payola's profits after tax before arriving at the amount of profit that belongs to Payola's ordinary shareholders (remember the principle — the ordinary shareholders have a right to whatever remains after the company has met all its other obligations). So "equity earnings" or "earnings attributable to ordinary shareholders" will be calculated on the figure that remains after tax and any minority interests or preference dividends have been deducted.

A further complication: in the past you would sometimes have seen in a company's profit and loss account a figure for "exceptional items" or "extraordinary items". These were items that did not arise in the course of the company's normal trade. Suppose Payola happened to sell off a subsidiary company in the course of the year at a profit of £200,000. Selling companies was not part of Payola's normal business, and it was an item that was unlikely to crop up every year. It therefore needed to be shown separately from Payola's income from its normal business.

The rule used to be that exceptional items were added to (or knocked off from) profits before tax — "above the line" in the jargon. Extraordinary items were accounted for at the net level after tax — "below the line". The difference between the two was not always clear-cut — extraordinary items were meant to be the very unusual items that were furthest removed from the company's normal business and least likely to crop up frequently.

However, the accounting authorities decided (rightly) that extraordinary items were being abused. One-off losses tended to be treated as "extraordinary" and tucked away out of the spotlight below the line whereas one-off profits were probably classified as "exceptional" and therefore allowed to swell pre-tax profits. So "extraordinary items" have now been virtually abolished and all one-off items are treated as "exceptional" and therefore affect the pre-tax profit figure, which is the one most frequently quoted. Today's accounting standards also

require the profit and loss account to distinguish between profits from on-going businesses, profits from businesses acquired during the year, and profits from any businesses which were discontinued in the course of the year.

The important point for the analyst of company accounts is that he is trying to form a view of what the company is capable of earning on an ongoing basis. So if the company has added a large exceptional item before arriving at its pre-tax profits, the analyst will normally strip it out (and adjust the tax charge accordingly) before arriving at the figure on which he calculates earnings. Likewise, an exceptional deduction from profits will be added back in. Some companies are not above using exceptional items to try to present an inflated view of their performance.

Finally, there is the question of convertible stocks and their effect on future earnings. We saw earlier how an analyst could adjust asset values to take account of a convertible and come up with a figure for fully-diluted assets per share. A comparable operation can be undertaken for earnings.

Payola currently produces £300,000 of pre-tax profit and net profits of £195,000 after all deductions except dividends. It has 3m £1 ordinary shares in issue, and pays a dividend of £130,000 net. Its earnings per share are 6.5p and the net dividend per share is 4.33p.

Suppose it has in issue £2m of 7.5% convertible loan stock, on which the annual interest charge is therefore £150,000, which is deducted before arriving at pre-tax profits. And suppose (since we need an extreme case to illustrate the principle) that Payola's current share price is 240p but that the convertible stock was issued a considerable time ago when the shares stood at only 80p. The conversion terms are that each £1 nominal of convertible loan stock converts into one ordinary share of £1 — in other words, the conversion price is 100p and well below current levels. Thus, when the stock is converted it will result in the creation of 2m additional ordinary shares of £1.

What happens on conversion is this. The loan stock disappears as a liability of the company and therefore the company saves the £150,000 of annual interest it was paying. Its pre-tax profits therefore rise from £300,000 to £450,000 and net profits rise from £195,000 to £293,000 after 35% corporation tax.

But 2m new shares have been created and therefore the company's earnings have to be spread over 5m shares rather than the previous 3m shares. Earnings per share thus work out at 5.9p in place of the previous 6.5p, representing significant dilution of the company's earnings.

Even though the convertible stock may not yet be due for conversion, an investment analyst will tend to undertake this calculation — working out what the position would be if the stock were converted — to arrive at the "fully-diluted earnings per share" figure. Since the dividend a company pays or can pay normally depends on the profit it is capable of earning, a big prospective dilution of earnings per share is not good news for shareholders. It means that there is a considerable constraint on dividend growth. Even at the present dividend level of 4.33p net, the dividend on the new 5m shares would cost £216,500 as against £130,000 on the old 3m shares and if the company maintains its dividend at this level, dividend cover will reduce, at least temporarily. Convertibles are not always good value for companies with rapid earnings growth.

23
Untangling a property trader

The accounts we have looked at so far have been for a property investment company with a small income from trading. How do the accounts of a trading company differ?

The obvious difference is that profits on the sale of properties will account for a far higher proportion of income. The company's business is mainly to develop a property, let it and sell the completed investment at a profit. It may also be trying to keep some of its developments as investments to build up a more secure rental income. And it will probably have some rents from properties awaiting development or developed and let but not yet sold. But trading profits will predominate.

The biggest problem in assessing a company of this type is that the profit figures for any one year do not actually tell you very much. Judgements have to be based on the management's success in generating trading profits over a period of years rather than in a single year. And the figures themselves can sometimes be confusing, particularly the preliminary profit figures issued some weeks before the full report and accounts are published. Companies have been put under pressure to improve their preliminary announcements, so with luck our example (see page 142) will be rarer in the future.

Floggem Properties — preliminary announcement

Profits for the year to March 31	Latest year £	Previous year £
RENTAL AND PROPERTY TRADING INCOME:	6,000,000	3,000,000
INCOME FROM ASSOCIATES	4,000,000	2,000,000
Less: INTEREST	2,000,000	1,000,000
PROFIT BEFORE TAX	8,000,000	4,000,000
Less: TAX	2,800,000	1,400,000
NET PROFITS	5,200,000	2,600,000
DIVIDENDS	2,000,000	1,000,000
RETAINED PROFIT	3,200,000	1,600,000

On the face of it, Floggem has done very well. Profits have doubled right down the line and press comment is quick to congratulate the company. The dividend is very well covered by available earnings. But we get a slightly different picture when we see the audited profit and loss account with the full report:

Floggem Properties

Consolidated profit and loss account for the year to March 31	Latest year £	Previous year £
Rental income	1,000,000	1,500,000
Property trading profits	3,000,000	1,500,000
Income from associates	4,000,000	2,000,000
	8,000,000	5,000,000
Add: EXCEPTIONAL ITEM: PROFIT ON SALE OF INVESTMENT PROPERTIES	2,000,000	Nil
TOTAL	10,000,000	5,000,000
Less: INTEREST	2,000,000	1,000,000
PROFIT BEFORE TAX	8,000,000	4,000,000
Less: TAX	2,800,000	1,400,000
NET PROFIT AVAILABLE	5,200,000	2,600,000
DIVIDENDS	2,000,000	1,000,000
RETAINED PROFIT	3,200,000	1,600,000

The first point is that a fair chunk of Floggem's profit — £2m before tax or £1.3m at the net level — comes from the sale of investment properties rather than those held as trading stock under current assets. This profit was lumped in with profits from other sources in the fairly rudimentary preliminary profits announcement on which most of the press comment was based. The full accounts show the real picture. Admittedly, the distinction between investment properties and trading properties can be a little artificial. But is the profit on sale of investment properties likely to be a recurring item? In other words, is the company generating further "investment" properties to sell in future years? Or is it simply selling the family silver?

Second, it is clear that the dividend is not covered by rental income, which is normally the most stable form of income, and the full accounts show that rental income has actually dropped over the year — perhaps because those rent-producing investment properties have been sold.

We also have a problem in calculating the pre-tax income that comes from rents because it is not always easy to see how the interest charges should be allocated between rental and trading income. A conservative approach is to assume they all represent the financing costs of the rent-producing properties. In any event, at £2m interest charges are well in excess of the £1m rents that the company produces.

In fact, if we look at the notes to the accounts we find that Floggem "capitalises" the interest charges relating to developments and to its trading properties: they are added to the capital cost of the projects in question and do not appear in the profit and loss account. Floggem's total interest charges in the year were actually £7m. The £5m that is not charged against profits has been added straight to the cost of trading properties. This is quite usual and should not in itself cause concern, provided we think the value of Floggem's projects will ultimately be higher than their all-in cost. But it does mean that, in the past year, the real outgoings by way of interest at £7m are almost equal to the group's total income of £8m ignoring the profits on the investment properties.

Next, we might look a little more closely at the £4m profit item that represents Floggem's share of the profits of its associated companies (companies in which it has a significant stake, but which it does not technically control). It is clear that much of Floggem's activity is carried out via joint-venture companies of this kind rather than by the parent itself or its subsidiaries. And the very large borrowings incurred by these joint ventures therefore remain "off balance sheet". They do not appear in Floggem's group accounts.

We might well discover with a little detective work that the profits of the joint-venture companies in question total £10m, of which Floggem's share is £4m. We might further discover that interest charges of £12m have been capitalised in the joint-venture companies and have not therefore been deducted in arriving at that £10m profit.

We might also find out that £6m of the associated companies' profits come from the sale of one particularly large devel-

opment and the remaining £4m from the sale of a plot of residential development land in the Toytown area of London's docklands, which the associate company in question had been sitting on for some years. There are no other similar plots of land to sell in the future, nor are there any further major developments in the associates likely to be sold in the current year. So the £10m profits of the associated companies (and Floggem's £4m share of them) look very much a one-off item. They are not a sustainable source of income. Moreover, that £4m profit share may in practice have been kept in the associates rather than paid to Floggem by way of dividend.

By now Floggem's originally declared £8m pre-tax profit for its latest year takes on a very different aspect. Rents do not cover the cost of the dividend. One-off sales of investment properties and of the associates' assets have in any case inflated the figure. In the absence of evidence to the contrary, profits are likely to be quite a bit lower in the current year. And the profits are struck, both in the Floggem group itself and in its associated companies, before charging massive amounts of interest that do not appear in the revenue account.

What is Floggem up to? Were we sufficiently cynical we might suspect that it has chosen to put every available item of profit in the shop window this year in order to get its share price up. Why? Perhaps it is thinking of making a rights issue of new shares for cash to help reduce those horrendous interest charges. Perhaps it is planning to use its shares as takeover currency to acquire another property group with more substance behind it. Time will tell.

Many property trading companies do, for a considerable period of years, produce substantial and growing profits. We are not suggesting that Floggem is typical of the sector. But our examination of its results brings out some of the points on which an investment analyst should focus.

24
Off balance sheet traders

We have seen that reading the accounts of a property trading company can be a lot more difficult than reading those of an investment company. With the investment concern, the bulk of the income comes from rents and we can see the value of the properties that produce the rents. The trader is deriving much of its revenue from the sale of properties and we probably will not know in advance what is likely to be sold in a particular year and at what profit. Interpreting the balance sheet can pose particular problems.

The modern property development and trading company may well undertake much of its activity through associated companies or various other forms of connected company whose accounts are not consolidated with those of the parent in the group balance sheet (see also Chapters 18 and 37). These companies probably borrow the finance they need for their developments on a non-recourse or a limited-recourse basis.

Thus the lender's security for the loan is the particular development for which the funds are being used, and the loan is not guaranteed (or not fully guaranteed) by the parent company. When this is the case, the consolidated balance sheet is unlikely to give a very helpful view of the real scale of the company's operations (or its commitments), and the notes to the accounts may be a lot more revealing than the accounts themselves.

Look at Floggem Properties. The investment properties are relatively insignificant at £10m (some of them were sold during the year, so the value has actually decreased). The trading properties — the developments held as current assets — are a more significant £30m. But these are the properties that Floggem itself or one of its subsidiaries is developing. In practice, most of Floggem's activity is channelled through various forms of related company which appear under the heading of "investments" in the consolidated balance sheet. And what appears in Floggem's accounts is the value of its share of the net assets of these related companies. Their gross assets are probably many times larger.

Floggem Properties — *Consolidated balance sheet for year to March 31*

	Latest year £m		Previous year £m	
INVESTMENT PROPERTIES		10.0		13.0
INVESTMENTS (associated companies)		25.0		20.0
		35.0		33.0
CURRENT ASSETS				
Developments in progress	30.0		25.0	
Debtors	5.0		4.0	
Cash	2.0		1.0	
	37.0		30.0	
CREDITORS (due within one year)	21.0		16.0	
NET CURRENT ASSETS		16.0		14.0
TOTAL ASSETS LESS CURRENT LIABILITIES		51.0		47.0
CREDITORS (due in more than one year)		17.0		16.0
NET ASSETS		34.0		31.0
SHARE CAPITAL (In 25p Ordinary shares)		5.0		5.0
RESERVES		29.0		26.0
SHAREHOLDERS' FUNDS		34.0		31.0

Take an extreme case. The associated companies in which Floggem is interested have gross assets of £200m, representing expenditure on a variety of development schemes, held as trading assets. The borrowings of these associates — in the form of non-recourse or limited-recourse loans — are a massive £137.5m to leave them with total net assets of £62.5m. Floggem's own stake in these companies averages out at 40%, so its share of the net assets is the £25m shown in its accounts.

Floggem's own borrowings are relatively small: £15m of short-term money included under the heading of "creditors (due within one year)" and a £17m term loan shown as "creditors (due in more than one year)". The really important group debt is the £137.5m borrowed by the related companies and the consolidated balance sheet itself would give us no inkling of this debt mountain. The interest charges do not go through Floggem's profit and loss account, and in fact are capitalised in the companies which incur them. (Fortunately, under new accounting rules, companies will in future have to give far more detail on associates and their borrowings in the notes to the accounts.)

Since Floggem has not made itself responsible for the borrowings of the related companies, which are secured purely on the projects they are undertaking, in one sense it is reasonable that they should not be reflected in Floggem's own accounts. On the other hand, if one of the developments in these companies went wrong, and the lenders looked like losing their money, Floggem would not be very popular if it simply walked away from the affair and left the bankers to bear

their losses. It might find it very difficult to raise further finance in the future. There is perhaps a moral or a practical commitment, if not a legal one.

In the late 1980s, much ingenuity was deployed in devising ways of keeping borrowings off the balance sheet: so-called "off balance sheet" finance. Part of the idea was to allow the company or group to appear relatively low geared, which can be reassuring to shareholders and lenders alike. The process was not confined to property companies, but since they traditionally borrow a large proportion of the funds they need, it surfaced in its most acute form in property company accounting.

It was, for example, possible to borrow money which, via a chain of transactions, would ultimately appear in the group accounts as preference capital rather than debt. Needless to say, any banker or investment analyst worth his salt will attempt to get at the commercial realities behind the accounts rather than satisfying himself with the bare figures as presented. And under recent accounting changes (see Chapter 16) companies themselves will have to give a far clearer picture of the commercial realities underlying their transactions.

Look carefully at the notes to the accounts for whatever information they give on the assets and liabilities of associated companies, non-consolidated subsidiaries and other companies or partnerships in which the group has an interest. The level of disclosure can vary quite widely. And look particularly at the item "contingent liabilities". This is where you should find details of any guarantees the company has given for the borrowings of non-consolidated companies or details of other legal liabilities the company may have incurred and which are not reflected in the balance sheet.

While Floggem, as a trader, will probably be valued by the stock market more in relation to its earnings than its assets, the asset figure is also relevant, at least as a fallback value. On the evidence of the published accounts, the asset value is 170p for each of the 20m shares in issue. But this is on the basis that the properties held as trading assets in the group and its associated companies are shown at cost. The real value could be quite a bit higher (it needs to be if Floggem is to show profits from property trading in future).

The chairman's statement might include some estimate of the likely current asset value, though make sure you note on what assumptions it is based. Given the gearing in the associated companies, every 1% appreciation in the value of their properties is worth 4p on Floggem's own asset value per share. But gearing can work in reverse, too, and problems in the associated companies could soon erode Floggem's asset value. This actually happened in the case of some very large property trading companies in the 1990-93 property market collapse.

In any case, unless Floggem or its associates are able to hold on to some of their completed developments as investments, any surpluses over cost in the trading properties will be coming through to the profit and loss account as income when the properties are sold. Tax will take its bite and part of the profit will be distributed as dividends. But an estimate of the surplus over cost in the developments does give some indication of the profit Floggem might be able to realise in the next few years.

25
Cash flow before profit

By the autumn of 1991 about a dozen property companies whose shares were formerly traded on the Stock Exchange had gone bust in the commercial property market's worst collapse in living memory. Many more were to succumb. In addition, many private developers had been wound up. In almost all cases they went bust because they ran out of cash to pay their bills — especially their interest bills.

Most of the comment in the press on property companies focuses on their published profits and asset values. And it is mainly the profit aspect that we have looked at in earlier chapters. But at a time of trouble in the property market, cash flow becomes far more important than profits. The difference between the two is all too frequently ignored.

To see the importance of cash flow, take a simple example. A developer constructs an office block at an all-in cost of £10m, which includes the interest charges on the £8m he has borrowed to cover the bulk of the cost. The building is worth £12m on completion, so the developer has created value — a "profit" — of £2m. But so far the cash flows have all been outwards: the £10m which has been spent on the development. And he has to repay the £8m of construction finance.

At this point he can do one of two things. He can sell the building for £12m to crystallise his profit, in which case he has ample funds (even after paying whatever tax is due) to repay his construction loans. This is what the trading company would do. Or, if the developer is an investor, he can negotiate a longer term loan on the property, which also gives him the cash to repay the construction finance.

This is how it should work in "normal" times for the property market. But, during a severe slump, a number of things happen. First, it becomes difficult to raise loans on property to replace the construction finance. Second, it also becomes difficult to find a buyer for the property and, third, the value of the property probably falls well below the £12m which it would command in normal times. And all these problems are greatly exacerbated if the property is not bringing in any income because the developer has not yet managed to find a tenant.

If the building remains untenanted, the position is this. The operation has already involved an outlay of £10m with nothing coming in. There is still £8m owing to the banks which cannot be repaid because the developer cannot sell the building at any reasonable price or raise further borrowings against it. And, for every year this situation persists, the £8m of construction finance from the banks is accumulating an extra £1m, say, of interest and the developer has no income to cover it. After a couple of years, if the banks have not closed in already, he could easily find himself owing more than £10m against a building which is now worth only, say, £8m. Not only does he have no cash to meet interest payments and loan repayment but — unless he has other assets — his liabilities now significantly exceed the value of what he owns — and the gap is widening by the day. He is well and truly bust.

But, to emphasise the importance of cash flow, imagine that this same development was being undertaken by an established property investment group with income from rents on its existing properties of £5m net.

This established company has plenty of income to pay the £1m interest charges on the £8m of development loans on the new offices. Provided they are getting their interest, the banks will probably be happy to extend their £8m of loans on the development for a few more years. The established company sees its available income reduced from £5m to £4m while the new property remains untenanted, and it suffers some fall in asset value because of the fall in value of the new property (and probably of its existing tenanted properties as well). But it is not in serious trouble. Because it has adequate cash flow it can afford to wait until the building finds a tenant and until the property market cycle turns up again, when values recover. It is exactly the same development as that undertaken by the trader. But the outcome is different because of the different cash-flow position of the two owners.

And, if we look at what happened to property companies during the early 1990s, we find that this phenomenon is clearly mirrored in the experiences of the different kinds of group. The developer-traders — companies which rely on sales of completed developments to provide their cash flow and profits — were mainly the ones which went bust. The established developer-investor groups, with a good cash flow from their existing properties, suffered a fall in asset values and in some cases a reduction in income, but they were not generally in serious trouble. Yet the quality of developments undertaken may have been much the same.

To see how misleading a profit and loss account can be as to the cash flow of a company, take the following hypothetical example. The accounts are those of another mythical property development and trading company called (perhaps optimistically) Sellitquik plc, which conducts much of its development activity through associated companies in which it has a large, but not a controlling, stake. Its operations are quite similar to those of Floggem which we looked at earlier. But it is a different aspect we are concerned with now.

Sellitquik plc — *Profit and loss account*	£m
Rents receivable	1
PROFITS ON SALE OF PROPERTIES	4
SHARE OF PROFITS OF ASSOCIATED COMPANIES	6
INTEREST RECEIVABLE FROM ASSOCIATED COMPANIES	3
	14
ADMINISTRATION EXPENSES	(2)
INTEREST PAYABLE	(5)
PROFIT BEFORE TAX	7

The great mistake here would be to equate the £7m of pretax profit with a cash inflow of £7m. In reality the £1m of rents probably represents mainly cash received, as does the £4m profit on sale of properties. In fact, Sellitquik sold properties with a book value of £12m for £16m cash, but at the same time it spent a further £20m on developing other properties for sale in future years.

The share of profits of associated companies requires closer scrutiny. These associated companies borrow money to undertake developments, and most of the interest on these borrowings is "capitalised": treated as part of the cost of the property rather than charged against profits. If it were charged against profits, these companies would be loss-making and Sellitquik would show in its own accounts a share of a loss rather than of a profit. In any case, the £6m which it shows as Sellitquik's share of the profits of associated companies is not received in cash (or in any other way, come to that). The notes to the accounts show that the associates paid no dividends at all to their owners.

The associated companies borrow direct from bankers, but they also borrow heavily from their parent companies. This is why Sellitquik shows interest of £3m on these loans, receivable from the associates. But note the word "receivable". The associated companies are as short of cash as Sellitquik and they do not pay interest on the loans from their owners. The interest is simply "rolled up".

Thus, of Sellitquik's total income of £14m, only £5m is received in cash. However, its administration costs of £2m have been paid in cash, as has the £5m interest shown in the profit and loss account on its own direct borrowings. But if you look at the notes to the accounts you would see that total interest charges were actually £12m, but the remaining £7m was simply added to the cost of properties. It probably had to be paid in cash all the same.

So, on revenue account alone, Sellitquik has only £5m of cash coming in, while it has £7m of cash going out and probably £14m if you include the interest capitalised. Even on this evidence it would be very badly placed to weather a property market downturn when properties became difficult to sell or borrow against.

Thus, the difference between profit and cash flow in a property company is vital. It is quite common for the latest accounts of a property company or a company with large property interests to show a "profit" right up to the time that receivers or administrators are called in because the company has run out of cash needed for continued trading.

Fortunately, with the arrival of cash flow statements in

published accounts it has become a lot easier to judge a company's cash flow position (see below).

The vital point to grasp when looking at the accounts of any property company is that "profit" is merely an accounting concept. There is no absolute definition of what constitutes a profit and the figure for "profit" shown in a property company's accounts can depend on a number of arbitrary decisions. Chief among these is the allocation of interest costs between what is charged against the year's revenue and what is "capitalised" by being added to the cost of properties.

Fortuitously, an event towards the end of 1991, while the property market was in free fall, illustrated very clearly the point we are making. So we will depart on this occasion from our normal practice of inventing hypothetical companies and turn to a real-life example.

Late in 1991 one of the major property development and trading companies of the 1980s, Stanhope Properties, was beginning to hit problems in the new climate of the recession. In consequence it was obliged to change its accounting policies in a way that underlined the risks of regarding published profits as indicative of a company's cash flow.

This is what happened. Stanhope and another property development and trading company, Rosehaugh, were equal partners in the joint company Rosehaugh Stanhope Developments (RSD), which had undertaken the major Broadgate office development in the City of London. As a 50:50 joint venture, RSD was not consolidated in the balance sheet of either parent. The interests of the two parents in RSD were shown in their own accounts simply as the value of their share of the net assets of the joint company, plus the value of loans which they had made to the joint venture. Both Stanhope and Rosehaugh operated extensively through "off balance sheet" companies, of which RSD was only one.

RSD raised loans on its own account for property development, but it also borrowed from its parents, as did various other of the joint ventures which each had. In its accounts for 1989-90 Rosehaugh included in its income some £23.5m of "interest receivable from related companies" on these loans ("related companies" would include "associates" or "joint-venture companies"). For 1989-90 Stanhope included, in its income, a sum of £17.4m as interest "receivable from associated undertakings".

The point to note in both cases is the word "receivable". The fact that interest on loans to associated companies was included as part of the income of the two parents did not mean that it was necessarily received in cash. In fact, it is now clear that much of it was not, but simply went to swell the figure for "debtors" — money owing to the company — in the balance sheets of the parent groups. For Rosehaugh and Stanhope decided in October 1991 to "waive" accumulated interest of £84.9m owing to them on loans they had made to RSD.

A fair proportion of this interest related to the period up to June 1990, meaning that interest had now been written off which had already been included in the profits of the parents in earlier years, though had never been received. This retrospective adjustment meant that profits of the parents for those years were, in reality, significantly lower than those published at the time (or losses were higher than published at the time). It was a very clear example of the difference between an accounting concept of profit and an actual cash receipt, and of the dangers of relying too heavily on the published figures for profits or losses in any given year.

While a company's cash flow can often be deduced from the published accounts and the notes that go with them, in the 1980s it had taken a fair bit of expertise to do it accurately. That has changed in the 1990s. Companies traded on the stock market, and some others, were required by the Accounting Standards Board to produce a "cash flow statement" for any financial period ending on or after March 23 1992.

The new accounting standard made it a great deal easier to monitor a company's cash-generating ability (or lack of it!) and proved a very useful supplement to the profit and loss account. In the light of experience, certain weaknesses were identified, however, and a revised version of the standard was published in the autumn of 1996.

Our example of a hypothetical cash flow statement for XYZ Group plc is taken from the appendix to the new standard, and it is worth looking at what the different headings would cover. While not specifically geared to a property company, it may easily be seen in property company terms.

EXAMPLE: XYZ GROUP PLC
CASH FLOW STATEMENT FOR THE YEAR ENDED 31 DECEMBER 1996

	£000	£000
Cash flow from operating activities (note 1)		16,022
Returns on investments and servicing of finance (note 2)		(2,239)
Taxation		(2,887)
Capital expenditure and financial investment (note 2)		(865)
Acquisitions and disposals (note 2)		(17,824)
Equity dividends paid		(2,606)
Cash outflow before use of liquid resources and financing		**(10,399)**
Management of liquid resources (note 2)		700
Financing (note 2) — Issue of shares	600	
— Increase in debt	2,347	
		2,947
Decrease in cash in the period		**(6,752)**
Reconciliation of net cash flow to movement in net debt (note 3)		
Decrease in cash in the period	**(6,752)**	
Cash inflow from increase in debt and lease financing	(2,347)	
Cash inflow from decrease in liquid resources	(700)	
Change in net debt resulting from cash flows		(9,799)
Loans and finance leases acquired with subsidiary		(3,817)
New finance leases		(2,845)
Translation difference		643
Movement in net debt in the period		**(15,818)**
Net debt at 1.1.96		**(15,215)**
Net debt at 31.12.96		**(31,033)**

The first item is a cash inflow of £16.022m from operating activities. This is the cash that the company generated from its operating activities over the year and it is unlikely to correspond exactly with the profit shown in the profit and loss account. In the case of a property investment company the operating cash flow would consist mainly of rental income received, less rents paid and the costs of running the business.

Adjustments must be made to the profit reported in the profit and loss account for items such as depreciation (which does not involve cash movements) and for increases or decreases in stocks and debtors. The details of the adjustments necessary would be provided in one of a series of notes to the cash flow statement — to avoid confusing with an excess of information, we have not reproduced the notes here. In the case of this example, however, one point appears very clearly from the detail in the note. In its profit and loss account the company had taken credit for its share of the operating profit of an associated company, amounting to £1.420m. But the dividend received in cash from this associate was only £0.35m. Therefore the £1.07m difference between the two has to be deducted from reported profits in arriving at cash flow.

The heading "Returns on investments and servicing of finance"would include interest received (not "receivable", note) and interest paid (not "payable"). Any preference dividends paid during the year would also be shown here. Interest paid could include interest that was capitalised (added to the cost of properties) during the year. The cash flow statement may thus be a better guide to the company's true interest burden than the profit and loss account. Again, the details of all these items will be given in a note.

The "Taxation" heading is pretty self-explanatory, though note that the corporation tax actually paid during the year probably does not correspond with the corporation tax liability that accrued during the year, which is what the profit and loss account figures will be based on.

"Capital expenditure and financial investment", at £0.865m, is a comparatively small item in this example. It would probably be a lot bigger in the case of a property investment company that was an active developer, because this is where expenditure on developing properties or acquiring additional properties would appear. Proceeds from the sale of investment properties would also come under this heading.

The heading "acquisitions and disposals" refers to acquisition or sale of subsidiary companies or whole businesses —

plus stakes in joint-venture companies — rather than transactions in individual properties. If the example company bought another property-owning company, the cash expenditure involved would show under this heading. Note that any debt acquired with the new subsidiary would also show up here. On the other hand, if the purchase was made for shares (or partly for shares, as is the case in this example) the share element in the purchase consideration would not be recorded here as it does not involve flows of cash.

"Equity dividends paid" is again largely self-explanatory: the cash cost of the dividends paid during the year on the ordinary share capital. Again, this may not correspond with the figure shown in the profit and loss account for the cost of dividends, since this latter figure would include the final dividend proposed but not yet paid out at the year-end. Dividends actually paid during the year will probably be the final payment for the previous year plus the interim payment for the latest one.

After all of these items we arrive at the cash flow position from the company's operations over the year. In this case the result is a cash outflow of £10.399m.

But this figure needs to be seen in context. It is clear that the company generated more than enough cash to cover its "revenue" outgoings: interest costs, tax payments and dividends. This is the first point to check on — there is reason to be very cautious about a company that does not generate enough cash to cover its "running" costs.

But the company in the example clearly expanded its business very significantly during the year — in this case via a major acquisition. The large outlays on acquiring additional businesses absorbed more cash than the company generated.

Turning some investments into cash raised £0.7m: this is shown under the heading of "Management of liquid resources". But even after £0.6m raised by a small issue of shares and £2.347m coming from new borrowings (less loan repayments), cash decreased by £6.752m over the period.

The second half of the cash flow statement provides a "Reconciliation of net cash flow to movement in net debt". As the heading suggests, this demonstrates the interaction between, on the one hand, the company's generation and absorption of cash over the year and, on the other, the net debt shown in the year-end accounts.

If we return to our hypothetical property trader, Sellitquik, we can see the relevance of the cash flow statement. The £6m share of profits of associated companies and the £3m of interest receivable from associated companies that featured as "profit" in Sellitquik's profit and loss account would not appear in the cash flow statement unless they were actually received in cash during the year. On the other hand, the £7m of interest capitalised would appear as an outgoing if it was paid in cash.

Profits of associates would feature in the cash flow statement only to the extent that they were paid to the owners by way of dividend — which in the case of Sellitquik they were not. The major differences between Sellitquik's profit and loss account and its cash flow statement would cause any analyst to look a bit more closely at the quality of its "profits".

26
Launches on the stock market

The most detail you are ever likely to receive on a property company is when it first launches on the stock market. Any company marketing its shares to the public needs to produce a prospectus which contains information required under company law plus quite a lot of additional information to satisfy the Stock Exchange's own requirements. If you are thinking of applying for shares, the prospectus should be read in detail.

But first there are a few general points to appreciate about stock market launches. A company can arrange to have its shares traded on the stock market in one of several ways. An "introduction" is when it already has a large number of shareholders — probably several hundreds — and is not seeking to raise fresh capital at the time. It simply seeks permission for the shares to be dealt in on the market. This is not immediately of great interest to the general public.

The second method is a "placing". Shares are sold privately to a range of investors — normally clients of the broker sponsoring the issue — and permission for them to be traded on the market is obtained at the same time. Only investors who are clients of the relevant broker are likely to be involved at this point. A variation on the placing theme is the "intermediaries offer" where shares are offered to financial intermediaries such as investment banks for them to allocate to their clients.

The third method, which gives the public at large an opportunity to apply for shares, is the "offer for sale". A prospectus and application form (or an invitation to apply for them) are published in newspapers and available from brokers or banks.

Sometimes, various combinations of offer for sale, placing and intermediaries offer are used.

Normally, in the case of an offer for sale, the shares are offered at a fixed price, which is calculated by the sponsoring broker or bank by reference to the prices for shares of comparable companies which are already traded on the market. The price is usually pitched a little below the likely value, to encourage applications with the bait of an immediate paper profit when dealings on the market begin. It also gets a company off to a good start on its stock market life if the issue is oversubscribed (there are applications for more shares than are on offer).

Companies could until recently launch either on the Stock Exchange itself (the "main market", in which case they are described as being "listed"), or on the junior Unlisted Securities Market or USM (in which case the shares are generally described as being "quoted" on the USM). Trading procedures are much the same in the two markets, both of which are run under the Stock Exchange aegis. The USM is, however, being disbanded. A new lower-tier market — the Alternative Investment Market or AIM — was launched in the summer of 1995.

The main difference between the main market and the USM was that a three-year profit record was normally required for the main market whereas two years had been acceptable for the USM. It was also possible to float a smaller amount of the capital on the USM than the main market, and flotation costs could have been somewhat lower.

The first point to look for when shares are offered for sale to the public is whether the shares on offer are new shares created by the company to raise further finance or whether they are simply existing shares being offered for sale by the present shareholders. In the second case the proceeds of the issue go to these shareholders, not into the company.

It is perfectly reasonable for existing shareholders to want to turn part of their stake into cash once they have built the company to a size to come to the market. But if all of the shares come from this source and nothing is being raised for the expansion of the company, there may sometimes be a suspicion that founding shareholders are selling out on the crest of a wave, which is not necessarily a reassuring pointer to the company's future prospects. In practice the shares on offer are often a mix of shares from existing holders and new shares being sold for the benefit of the company itself.

The second point to look for, before you even begin to consider the company's profit record and prospects or the price being asked for the shares, is the nature of the business. Suppose Floater Properties Plc is launching on the market. Is it primarily a property investment company? Is it a development company? If it is a developer, does it sell its developments on completion (in which case it is a "trader") or does it retain them for investment? Or is it simply a dealer, making profits from buying and selling land and buildings?

The prospectus will usually contain a "Key information" section giving a quick pointer to the type of business, and this business will be described in more detail in the body of the prospectus.

The distinction between different types of property company is important for reasons we have already examined. If Floater is an investment company (probably also undertaking developments which it will retain as investments on completion), its income is likely to come from relatively safe rents which will probably increase over the years. This is high-quality income. The shares in this case are likely to be valued mainly by reference to the asset value.

Since the shares of existing property investment companies on the market probably stand at a discount to their asset backing, Floater would need to be launched at a discount, too. If the asset backing was 100p per share, the shares might be offered for sale at, say, 70p or 75p.

However, it is not just the current asset backing that is relevant. If Floater is undertaking major developments that will throw up a big surplus over cost when they are completed in the near future, these surpluses will shortly raise the asset value substantially. So investors need to look at prospective asset backing as well as the figure disclosed by the latest accounts.

In practice, launches of pure investment companies are relatively rare nowadays. This is partly because the entrepreneurs who have built up a substantial private property investment company may see little point in a market quote which is likely to value £1 worth of assets at only 75p or so, thanks to the discount factor. There was, however, a spate of investment company launches early in 1994 when property company share prices briefly rose to stand at a premium to asset value.

If Floater is a property trading company, the investor has to be rather more careful. Most of the income in this case will come from profits on sale of properties. It is of a lower quality than rental income, partly because there is not the same assurance that it will prove sustainable. And the shares will be valued mainly on an earnings basis — by reference to a price earnings (PE) ratio rather than to an asset value. The price asked for the shares in this case may be well above the asset value rather than below it.

This is fine so long as the company can continue to generate increasing earnings from developing buildings and selling them at a profit. Suppose Floater has assets of 30p per share and reckons to generate earnings of 9p net per share this year. The shares might be offered for sale on a PE ratio of 10 — in other words, at 10 times earnings or 90p each. But if the property market should turn sticky and Floater finds it can no longer generate the profits, the only ultimate backstop for the share price is the asset value. Valued on an assets basis and without the high earnings, Floater might be worth little more than 20p in the market. For the investor there is plenty of "downside potential".

It is easy to see why property trading companies like to launch on the stock market — selling 30p of assets for 90p is good business. But developing and selling property is rather different from other forms of manufacturing activity. A producer of baked beans, say, may build up customer contacts and brand loyalty and can expect a reasonable demand for his product from one year to the next. It is difficult for a property trader to build up goodwill in quite the same way. His market is a lot less certain and much more likely to dry up from time to time, and his livelihood is correspondingly more precarious. Be very wary of property trading companies that launch on the back of a profit record of only a few years. It may have been earned in a boom time for property, when it would have been difficult not to make money.

Once into the nitty-gritty of a property company prospectus, one of the first items we hit is a brief history of the company. This can be useful. It helps to give a feel for the way the company developed and what it now is. Suppose it was formed about 12 years ago. Is it still run by much the same management team? Has it radically changed the nature of its activities over the period? Has it taken over other companies during the 12 years, or has it simply grown organically?

These points are important, because they help to tell you how much weight you should give to the profit record. If the company has been doing much the same thing for 12 years (though presumably on an expanding scale), it will have had to deal with tough times as well as the good ones. The fact that it has survived provides some reassurance as to its ability to cope with problems in the future.

If, on the other hand, the company was effectively a housebuilder until three years ago, then moved into commercial property developments and made good profits on two or three lucky deals, the recent profit record will be more suspect as a guide to long-term potential.

Even if the company has always been a commercial property developer, but has vastly expanded the scale and range of its developments over the past three years, you may need to be a little cautious. It is very easy for companies to over-extend themselves by expanding rapidly in boom times. They are then far more vulnerable to a downturn in the market than if they were still developing on a small scale in areas they know well.

There was one property company that launched on the USM in the 1980s, with the requisite profit record, which had virtually changed the whole nature of its business over the previous 12 months, selling most of its existing properties and taking on a new and different portfolio. The historical record thus told you virtually nothing about its prospects. It was not, incidentally, a great success.

The next point to remember is that personalities are more important in property than in many other types of business. Successful property companies with many millions of pounds in assets have been run by two or three people operating from a couple of back offices. What is important is the ability to spot a profitable deal. It only takes one man to spot a site, visualise the development opportunity it presents and negotiate a purchase (and possibly the finance for the scheme). Most of the other skills necessary for the project can be bought in from lawyers, accountants, architects, surveyors and construction companies.

There are, of course, very large property companies with their own development departments, management departments, letting departments and so on. But even here you may find that the success of the company ultimately still depends on the flair of one or two individuals.

So the people who run the company are important, and the prospectus provides a fair bit of information on their background and careers. Look particularly at the chief executive: often the person who is chairman or managing director (or sometimes both). This is probably the entrepreneur who founded the company and provides much of the flair. If you are in any doubt, look later in the prospectus at the details of the directors' own shareholdings. The founding entrepreneur is probably the person with the biggest chunk. The history section may also have given you some insight into the characters involved in the company's growth.

If you have contacts in the estate agency and property world, they will almost certainly know something about this entrepreneurial character if he has made his mark on the property scene. Find out what you can. But do not rely too heavily on what you read about up-and-coming property entrepreneurs in the non-specialist financial press. A good public relations firm, a good lunch and a quick whip round a couple of newly erected warehouses near the M1 motorway can convince many a budding journalist that he is meeting property's next Mr Big. The reality is usually rather different.

In assessing a company, you are looking for flair at the top. But remember that the flair that brings off a couple of good deals or pulls a fast one over a rival developer is not necessarily the same skill that successfully runs a company and its finances, year-in, year-out. Once you have assessed the entrepreneur, look at the back-up and particularly at the finance director. Has he the experience to run a quoted company? And, for added comfort, is there perhaps a figure from the banking world on the board as a non-executive director?

Just occasionally in a prospectus you will find some less than flattering details of a director's past — he has been a director of companies that are now in forced liquidation, has been dragged through the courts for putting in multiple applications for British Telecom shares, or the like. It does not necessarily mean he will be a bad director. Neither does it always reflect particularly well on his intelligence or judgement. The reason that these details go in is that the sponsors to the issue insist. If there is anything that is or might be material to the outlook for the company, it has to be declared at the prospectus stage. Normally the sponsors play safe by revealing any dirty washing.

With these preliminaries out of the way, you can look at the section of the prospectus that describes the company's business in detail. Assuming it is a developer, there are two main areas to focus on: the type of development and the way it is financed.

This section should give a pretty good feel for the operation. Does the company concentrate mainly on office, shop or industrial properties? Does it operate in one particular area or across the country as a whole? Does it undertake most of its developments itself or in partnership with others (in which case, are the partners widely known names)? Does it undertake a multiplicity of small schemes or a smaller number of very large ones? Does it perhaps have too many eggs in one basket? Or has it perhaps built up an impressive track record

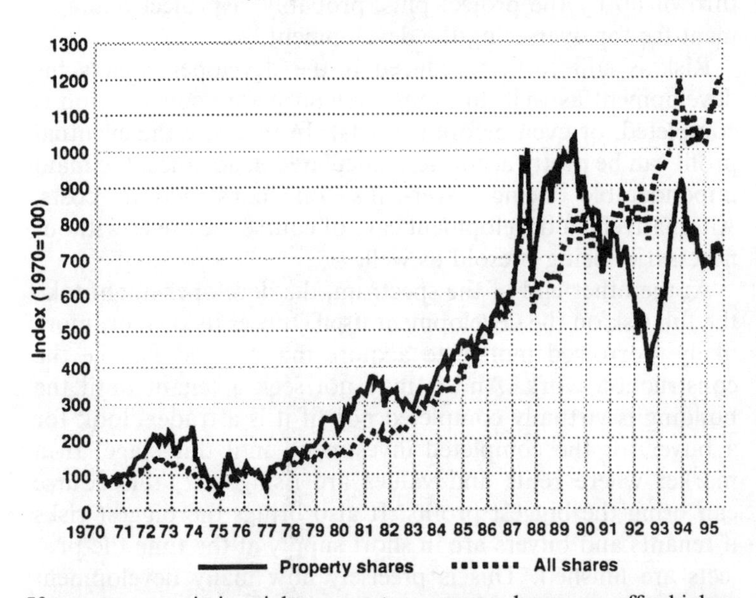

If you get your timing right, property company shares can offer higher rewards than shares in industrial and commercial companies as a whole. Get the timing wrong, and you can lose money in property shares much faster. The chart plots property shares (solid line) against the FTSE All-Share Index (dotted line), and clearly shows their greater volatility. Source: *Datastream*

in one particular development speciality: business parks or retail warehouses, for instance?

The section that describes the business should give information on the company's major properties or developments (greater detail will be given in the report by independent valuers, also included in the prospectus). An important point here: what are you told about the way projects are planned and financed? This may be the key to the degree of risk in the operation. .

If the company is a trader that "pre-funds" its developments, one element of risk is greatly reduced. Pre-funding normally means that the developer agrees in advance with an institution that the latter will put up the finance for the development and assume ownership on completion or sooner. The developer gets a profit related to the surplus over cost thrown up by the project plus, probably, a project management fee for overseeing the development.

Risk is still further reduced if the developer prelets his development as well (lines up a tenant before construction is completed, or even before it starts). In this case the eventual profit can be pretty accurately calculated in advance, the main imponderable being over-runs on construction costs. Alternatively, a development can, of course, be prelet without necessarily being presold as well.

At the other end of the spectrum, the developer might take the full risk on the development itself. It uses its own or, more likely, borrowed money to acquire the site and finance the construction work. And it may not seek a tenant until the building is virtually completed nor (if it is a trader) look for a buyer for the completed investment until this stage. In a market where rents and values are rising fast, this course can bring the biggest profits. It also brings the biggest risks if tenants and buyers are in short supply at the time the projects are finished. This is precisely how many development companies were caught out, when developments conceived in the boom days of the late 1980s became available for letting in the market slump of 1990-93.

There should also be some indication of the source of the group's profits, particularly if it undertakes a number of different activities: housebuilding as well as development of industrial estates, for example. Does any of the income come from rents? This would be a pointer that the company probably holds on to some of its developments for long-term investment, even if it is basically a trader.

Finally, what does the description of the business tell you about the company's ability (if it is a trader) to continue generating profits in the future? Are there a number of schemes in progress which will come through to profits in future years? What does the prospectus tell you about sites the company has acquired or arranged to acquire? This information needs to be interpreted with care. An ongoing development programme is a plus point. But you would be worried if it looked as if the group were biting off more than it could chew.

At this point you can begin to relate what you have picked up about the company's operations to the accounting and financial information provided. Look at what the company says about its reasons for launching on the stock market. There will probably be some pretty standard phrases about increasing the company's visibility, helping to attract development opportunities and high-calibre employees, and so on.

If a fair proportion of the issue proceeds goes to the company itself rather than its existing shareholders, there may be an indication that the finance is needed for site purchase, or to increase financing options. Sometimes a trading company will be increasing its financial resources so that it can hold on to some of its developments as investments in the future (additional equity capital provides a cushion which allows it to step up the amount of money it can borrow).

The company may make a forecast of profits for the coming year, though it is not obliged to do so. If it does, it will deliberately pitch the figure a bit on the low side. There is nothing worse for a newly launched company than to miss its profit forecast. It will probably also forecast the dividend it expects to pay. And it may produce a "*pro-forma*" statement of its assets, which shows the effect of including any money raised for the company by the issue of shares. This *pro-forma* statement also allows some updating of the position shown in the last published accounts — the figures for the company's latest year-end may already be some way out of date.

These forecasts for profits, dividend and asset value provide the basis for the calculations of PE ratio, yield and discount (or premium) to assets on the shares at the launch price. If based on forecasts, they will be "prospective" ratings. Stock market investors and share analysts will compare these ratings with the ratings of similar companies already established on the market and attempt to assess the attractions of the newcomer at the asking price.

Note that the forecasts, if there are any, are entirely the responsibility of the company's directors. There will be letters from the reporting accountants and banking advisers, saying that the calculations and accounting policies used are reasonable and that the company has made the necessary inquiries in arriving at its predictions. But these professional advisers make it quite plain that they are taking no responsibility for the company's ability to attain the figures it comes up with.

After the forecasts, look at the detailed profit record. This will appear in the report of the "reporting accountants", who are sometimes but not always also the company's auditors. The report will include the profit and loss account and the balance sheet for the past three or five years, usually heavily adjusted since the company will probably have produced the figures rather differently while it was still private.

Take the balance sheet first. Are properties held as fixed assets or as stocks included under current assets? In the second case the chances are the company is mainly a trader. Is there a breakdown of the properties (which, if held as trading stock, are almost certainly shown at cost)? This might divide the properties among completed developments awaiting sale, developments in progress and sites held for future development. It can help to show how much near-completed development will shortly be coming through to profits.

Does the company have investments in associated companies (which would suggest it may undertake some joint projects)? What further information is given on these associates (in particular, do they have heavy borrowings of their own which are not shown in the group accounts)?

What is the gearing of the group (the relationship of borrowed money to the shareholders' money in the business)? Ultra-high gearing may sound a warning note, but remember that the ratio will be reduced by whatever cash the issue is raising for the company. It is not unusual for a company to be highly geared ahead of the launch but to look a lot more healthy once the proceeds of the issue are in. Check whether the borrowings are long term or short term and whether they are at fixed or variable rates of interest (the notes to the accounts should help here). Short-term variable-rate borrowings offer the company a lot less security than longer-term fixed-interest ones.

Next, look at the profit and loss account. How does profit break down between rents and profits on sales of properties? Are there any other sources of income? Would rental income alone be sufficient (after tax) to cover the dividend the company is proposing, or is the payment reliant on trading profits? The picture you get from the actual accounts should confirm your earlier impressions of the way the company operates. Do profits include any one-off "exceptional" items which may have inflated the last year's figures but will not necessarily be repeated?

Above all, look at the way interest is treated in the accounts — you should be able to get the picture from the statement of accounting policies and the notes to the accounts. The interest charge shown in the profit and loss account is not necessarily the full amount the company has to find in a year. Part of the interest may simply be

"capitalised" — treated as part of the cost of developments, and added direct to the cost figures at which they stand in the books. While capitalising interest is perfectly acceptable practice, charging it all against revenue is obviously a more conservative policy. Earnings which have borne the full interest charge are of higher quality than those that have not.

A couple of final points from the notes to the accounts. You should be able to see the amount of future capital expenditure authorised by the board, which helps in assessing the scale of the group's operations. And the heading "contingent liabilities" should give you warning of any "nasties" lurking in the background — lawsuits, claims against the group for shoddy construction, etc — plus an indication of any bank guarantees given to joint-venture companies.

After the accounts, look at the property valuer's report — you will never again have as much detail on the portfolio. Not only will it contain a description of each property, tenure, main leases and the like, but, in the case of developments, it should give value in existing state, costs to completion and expected capital value when completed. It is worth mulling through these details to broaden your view of the type of operation the group undertakes.

If you have time, look also at the "general information" section that comes towards the end of the prospectus. Much of this will be fairly routine small print, including a summary of the main points from the company's Memorandum and Articles of Association (effectively, its written constitution). But there are usually some items of interest, such as the company's borrowing limits, the directors' salaries and service agreements, directors' shareholdings, incentive schemes and the like. Occasionally there will be information on lawsuits the company is involved in or on interesting contracts between the company and one or more of the directors.

Finally, comes the assessment. Are the shares worth applying for at the offer for sale price? The assessment has to be divided into three parts:

- Is it a good company?
- Are the shares good value at the offer price?
- Will other investors consider the shares worth applying for?

In the very short run, the last assessment may be the most significant. If investors think there are likely to be applications for more shares than are on offer, and that the price in the market will go above the subscription price, they will be encouraged to apply, thus adding to the likely oversubscription and the probable premium when dealings begin. This may have far more to do with fashion, market mood and a quick profit than with the real merits of the company. For long-term investors, the first two points are more important.

27 Property company takeovers

There's nothing quite like takeovers to set the property share sector of the stock market buzzing. For tax and other reasons, property investment company shares normally stand in the market at a discount to their estimated current net asset value, though this discount will be smaller in a booming market than in a dull one. But investors tend to assume — sometimes rightly, sometimes wrongly — that if a takeover is to succeed the bidder will need to offer a price close if not equal to the victim's net asset value. Why should investors be expected to sell £1 worth of assets for 80p?

The concentration on asset values makes property company takeovers rather different from those of industrial companies. But the same basic rules apply. If thrusting Payola Properties wants to acquire sleepy Dormant Estates, it has to persuade the Dormant shareholders to accept its bid. Let us suppose that Dormant's shares stand in the market at 160p ahead of the bid and Payola's shares are quoted at 80p. Let us suppose also that Dormant's last published asset value was 200p per share and Payola's was 100p. Payola can:

- Offer to swap its own shares for those of Dormant in the proportion of two-for-one. Two Payola shares are worth 160p, which is Dormant's current price. But investors would almost certainly expect more, so to win acceptance Payola might need to offer, say, five of its own shares for every two of Dormant. Five Payola shares are worth £4, so the offer values each Dormant share at £2 (which happens to be its asset value). This will, of course, have the effect of diluting

Payola's assets in the short term. If the bid were an agreed merger between the two companies, it is just possible that terms of two Payola shares for one of Dormant might be agreed (which reflects the assets contributed by each side). But unless the Dormant directors controlled the company with over 50% of the shares, the way would be open for another bidder to come in and offer a higher price. Dormant shareholders might accept this higher offer, even if their directors recommended against it.

- Payola could offer, say, 200p in cash per Dormant share. In this case the Dormant shareholders who accept have no further interest in the combined group after the merger. They simply take their money and run, paying whatever capital gains tax is due (capital gains tax is not payable at the time, however, if you simply swap your shares for those in another company).

- Payola could offer some other security of its own for the Dormant shares (Payola 7% convertible loan stock, say). Or it could offer a mix of convertible and shares. Or a mix of shares and cash. Or, if it made an all-share offer, it could offer a cash alternative as well (this is likely to be at a slightly lower price than the value of the share offer).

What the bidder offers depends partly on what it thinks will be acceptable to the Dormant shareholders. But it also depends on whether it wants to increase its gearing via the acquisition (in which case it will be more likely to offer cash or some form of loan notes). But these options will be expensive in terms of interest charges compared with the dividends on the shares it might offer instead. So it would have to be confident that it could increase gearing without damaging its earnings too severely.

If Payola has done its homework on Dormant, it might reckon there is scope for increasing the value of the Dormant property assets by active management: realising marriage values, exploiting development opportunities and so on. It may also reckon that, if it uses a more bullish set of external valuations than Dormant was doing, it will come up with higher figures. So it may not, in reality, be diluting its own assets via the takeover as much as appears.

Payola might also reckon it could make the earnings of the combined group look better with some accounting changes. A common trick would be to sell off some of the Dormant properties and treat any surplus as part of earnings in the profit and loss account.

But one thing Payola will have to watch out for is any potential tax liability in the Dormant portfolio. Suppose Dormant has properties with a value of £100m (most of which it has held for a long time) and net assets of £70m backing the 35m shares in issue (which gives the asset value of £2 per share). But Dormant reckons — and this would normally be shown in a note to its accounts — that if the properties were sold at their book values it would have to pay £15m of tax on the capital gain. It has not allowed for this £15m in the accounts themselves, because it has no intention of selling the properties.

In its calculations, Payola has to knock off this £15m potential liability from the £70m asset value, leaving net assets of £55m free of tax, equivalent to only 157p per Dormant share.

If Dormant resists the takeover bid and argues that the price is too low, Payola will make much of this tax liability in its counterblast.

Payola will need to gain just over 50% of the Dormant shares to be sure of control, though it will not need to go ahead with the bid unless it gets 90% (which is the level at which it can acquire the remaining shares compulsorily). Between 50% and 90% it will be up to Payola to decide whether it wants to proceed ("go unconditional") or not, except in the case of a "mandatory" bid — see below — where different rules apply. If it decides not to proceed, it simply lets the offer lapse and Dormant shareholders who have accepted get their share certificates back.

To help it to get up to the 50% control point, Payola may have bought Dormant shares in the stock market before it even launched its bid, but once it got above 3% of the Dormant capital (this level was lowered from 5% with effect from 31 May 1990) it would have had to declare its stake and investors might have guessed that there was a bid coming.

If it gets up to 15% this way there has to be a breathing space for Dormant to respond. And if Payola raises its stake to 30%, it has to bid for the rest of the Dormant shares if it has not done so already (a mandatory bid). It has to be careful in its stock market buying. There is a rule that says that, when it bids, it must normally offer a price at least equal to the highest price it paid for the shares over the previous year. But stock market purchases — which can take place once the bid is in progress as well as beforehand — can be a powerful weapon in winning a takeover battle.

The way a takeover proceeds will depend on whether the victim company is resisting the bid or not and whether there is more than one bidder. The job of the Dormant directors is to get the best possible deal for their shareholders (in the case of a hostile bid they are probably also sensitive to the fact that they may lose their own jobs if the bid goes through). So they may try to fight off the bid altogether. Or, if they reckon they cannot remain independent, they will hold out for the best possible price. To have a chance of winning, Payola may be forced to increase its terms one or more times during the course of the offer.

A main plank of Dormant's defence is likely to be its asset value. Ideally, a full independent valuation will be required, but if time is very short the Takeover Panel — which regulates these matters — may allow a partial revaluation or an "informal" valuation. If Dormant can show that its properties are worth £10m more than their £100m book value, this takes net assets up to £80m or 228p per share. It will also try to show its growth record in the best possible light. And it may attack the Payola record and prospects in an attempt to undermine the value of the Payola shares.

Meanwhile, Payola is likely to be attacking the performance of Dormant and its management. If it is making a share exchange offer, it is probably also relying on "friends" to buy its own shares to support their value and thus the price it is offering for Dormant. One of the oldest tricks is for a currently fashionable property trader, which has little in the way of assets but whose shares are riding high, to get some solid assets behind it by bidding in inflated shares for an asset-rich investment company.

As a final move, if Dormant really wants to keep out of Payola's clutches and sees no other way, it may look round for a "white knight". This is an alternative bidder who would be more acceptable to the Dormant directors and might also (though this will obviously not be spelled out) let them keep their jobs after the takeover.

28
Categories of finance

Having looked briefly at some of the principles of property financing and the way property company accounts are structured, we need to examine in more detail the different types of finance available. In the process we will be categorising property finance on a number of different criteria. Perhaps we should define these at the outset.

• **Debt or equity.** This is the most basic distinction of all. Is the money being borrowed, or is it put up on a risk-sharing and reward-sharing basis? Borrowed money — debt — has to be repaid at some point and interest (or the equivalent of interest) has to be paid on it until the debt is repaid. With equity, the return for the person who puts up the money is determined by the success of the enterprise. He shares in the profits and if there are no profits he gets no return.

The most obvious form of equity is money subscribed as ordinary share capital of a company. But there are many other equity financing arrangements. The provider of finance for a particular development may, for example, be entitled to a share of the profits even though the money is not put up in the form of share capital. And there are deferred forms of equity, like the convertible loan. This starts off as debt, paying a fixed rate of interest. Later it may be converted into shares of the company, at which point it is transmuted into equity.

• **Loan or traded security.** Borrowed money may be a simple loan. You borrow, say, £1m as an overdraft from the bank. You pay interest on the loan and at some point you pay back the £1m to the bank. Or you might instead borrow £1m (in practice it would only be worthwhile for larger sums) by inviting investors to subscribe for a bond of one sort or another. This is a form of IOU (I owe you) note, and is a security rather than a straight loan.

In return for providing the £1m, the investors get a piece of paper entitling them to receive such-and-such an amount of interest each year, and to have the money they put up repaid at a specific date. But they do not have to wait until this repayment date to get their money back. If they need the

money sooner they can sell the piece of paper to other investors, who then become entitled to receive the interest and the eventual repayment. Securities of this kind are normally bought and sold on the stock market or in the euromarkets.

• **Secured or unsecured.** Lenders normally want security for the money they provide. In other words, they want a charge on specific assets of the business or on the assets of the business as a whole. What this means is that, if the borrower fails to pay interest or make capital repayments on the due date, the lender can put a receiver into the enterprise; his job is to sell the assets that have been charged and repay the loan from the proceeds.

A loan secured in this way is clearly safer than an unsecured-loan. But well established companies may be able to raise unsecured loans because the safeguard for the lender is the company's established profit record, out of which the interest can comfortably be paid. Most borrowings in the euromarkets (see below) are unsecured, as are borrowings in the commercial paper market. The company's name and standing are the main guarantees.

• **Fixed-rate or variable-rate (floating-rate) interest.** With some borrowings the rate of interest is agreed at the outset and remains unchanged over the life of the loan. With others, the interest rate will change according to movements in interest rates in the economy at large. For large-scale floating-rate borrowings the most common yardstick of interest rates is the London Interbank Offered Rate, or LIBOR. This is the rate of interest at which the banks themselves are prepared to lend to each other (see next chapter for the detail).

Short-term borrowings are most likely to be at a floating rate of interest. Long-term borrowings such as mortgage debentures are more likely to be at a fixed rate.

But there is another complication we will meet. A company may make a one-off payment to buy a "cap" for a floating-rate borrowing. This is a form of insurance policy which means that, whatever happens to interest rates in general, a maximum limit is set to the interest rate that the borrower will have to pay. This and other methods of hedging the interest rate risk are discussed in Chapter 31.

• **Long term or short term.** You can borrow money for a year or you can borrow it for 30 years (if you are well enough established). One year is clearly short term and 30 years is clearly long term. In between, there are medium-term borrowings. Where does the dividing line fall? How long is a piece of string? But we have seen that, in company accounts, borrowings for less than a year are categorised separately from other borrowings.

However, we have to be a little careful in using these terms. An overdraft is technically repayable on demand, so it has to be categorised as a very short-term borrowing. But in practice many companies have an overdraft outstanding almost indefinitely. Likewise, under a multi-option facility (MOF) a company might technically be borrowing for three months at a time. But when the three months are up it simply repays the original loans and takes out new ones for another three months. If the facility runs for five years, effectively it has the use of five-year money via this "rolling over" arrangement (see Chapter 30).

• **Project finance or corporate finance.** A smallish or young company undertaking a property development probably does not have a great deal of security to offer other than the development itself. So it borrows money on the strength of this development. This is project finance. A longer-established and larger company may be able to borrow on the strength of the company itself rather than its individual projects. This will be corporate finance. Interest rates will generally be rather lower for corporate loans than for project loans.

But even some quite large companies are undertaking developments of such a size that they dwarf the company's own resources. In this case each development may need to be financed separately on a project finance basis and may well be "off balance sheet" (see below).

• **Recourse, non-recourse or limited-recourse loans.** If the loan for a particular project or for a particular subsidiary company is guaranteed by the parent, the lender has recourse to the parent (can require it to stump up) if the project itself gets into trouble. But if the only security for the lender is the project itself, and the parent company has given no guarantees, the loan is "non-recourse". In practice, many loans are "limited recourse". The parent may, say, have guaranteed interest payments but is under no obligation to repay the loan itself. Because the parent company has no liability or a very limited liability, non-recourse or limited-recourse loans to non-subsidiaries do not appear on the parent company's balance sheet. They are "off balance sheet". But remember that under the new accounting rules it will be far more difficult to disguise a subsidiary as something else. If the borrower qualifies as a "quasi-subsidiary", its debts will need to be included in those of the group.

• **Domestic or euro finance.** Finally, loans may be in sterling or in other currencies, and they may be raised through the domestic markets or the euromarket. The euromarket is the international market in "stateless" money, which is centered on London, but is not part of the domestic British financial system. Borrowings in the euromarket, which can be made in most major currencies including sterling, tap the very large deposits of different currencies held in banks worldwide. But companies generally need to be large and internationally recognised to follow the euro route.

29
Financing with commercial paper

In looking at the financing options available to property companies we will start with the very short end of the market. Whereas in the past companies generally satisfied their short-term needs by borrowing from the banks, nowadays they have greater opportunity to tap the money markets direct, often with a significant saving in cost. This is where the commercial paper market comes in.

The sterling commercial paper market got under way in 1986. Initially its use was restricted mainly to the larger stock market-listed companies, but the restrictions were considerably relaxed in the 1989 Budget, opening the way for medium-sized companies to tap this source of funds.

Commercial paper is a form of short-term — often very short-term — IOU (I owe you) note. The company that wants to borrow — the issuer — sells its commercial paper to investors looking for a short-term home for surplus cash. It is thus issuing a security, though a very brief-lived one, instead of taking up a loan. This is an example of the securitisation process at work. Commercial paper is normally sold at a discount rather than paying interest as such.

Thus if a company were issuing paper with a life of three months, it might sell it at, say, £97.5 for every £100 of nominal value. At the end of three months the paper is repaid at £100 per £100 of nominal value. Thus the investor has a profit of £2.50 per £100 nominal over the three months, which can be expressed as the equivalent of an annual rate of interest.

As with most money-market financing techniques, the "interest rate" is normally expressed in relation to LIBOR:

the London Interbank Offered Rate. This is the rate at which the banks themselves will lend to each other. The lower rate at which they are prepared to borrow is known as LIBID: London Interbank Bid Rate. And the middle rate between the two is LIMEAN (pronounced "lie mean"). Interest rates are usually quoted as being "so many basis points" above LIBOR or LIBID. A basis point is one hundredth of a percentage point, so 25 basis points over LIBOR would mean a quarter of a percentage point over LIBOR. If LIBOR were 10%, the rate of interest would be 10.25%. LIBOR itself is the constantly changing measure of the cost of funds to the banks themselves.

Financial institutions — and other companies — invest in commercial paper, as do the banks. Its maximum life used to be limited to a year, but after the 1989 Budget this was extended to five years. However, the average life of commercial paper is more like 40 days.

What use is money that has to be repaid in 40 days, or even in three months? The answer is that, though commercial paper normally has a short life, a commercial paper programme can be used as a medium-term source of finance. Paper is issued for, say, three months. When that paper falls due for repayment, a further issue can be made for another three months. And so on. This "rolling over" process is a familiar technique in the money markets and it allows short-term money from investors to be transmuted into longer-term financing for the borrower.

A company that wants to tap the commercial paper market needs first to appoint a bank to set up a programme for it. The size of the programme — £50m, say — is the maximum amount of paper that the company may have outstanding at any one time. It also needs to appoint a dealer or dealers; there will usually be more than one. These are banks which act as intermediaries between the issuer and potential investors. When the company wants to raise cash by issuing paper it informs its dealers who are constantly in touch with prospective investors. Sometimes it works the other way round and the dealers tell the company that there are investors looking for paper of such-and-such a maturity, and is the company interested in issuing?

The issuer also needs to appoint an issuing and paying agent who will look after the technicalities of getting the paper physically into the hands of the investors and redeeming it on the due date with funds supplied by the company. Even after the relaxations in the 1989 Budget, the minimum denomination for sterling commercial paper is £100,000, so the market is not for very small issuers or investors.

As commercial paper is unsecured, the standing of the borrower is very important. When the market was effectively restricted to large household-name companies it was not too difficult for investors to judge the credit risk. But now that medium-sized and possibly less familiar companies can participate, there is likely to be an increase in the proportion of issuers seeking a "rating" for their paper. A rating is an independent credit assessment, and the names Moody's and Standard and Poor's are well known for their rating services.

The credit rating helps investors unfamiliar with the money markets or more sophisticated investors considering buying the paper of companies with which they are not familiar.

The arranging bank will also undertake an education exercise among potential investors when the commercial paper programme is first set up. The company will probably produce an information memorandum about itself which can be circulated among investors and will help them to decide whether this is paper they would invest in, and up to what limits. Once a programme is established and investors are used to assessing the paper from the particular issuer, further issues can be made almost instantaneously.

Some companies tend to have a fairly constant amount of commercial paper outstanding, while others tap the market only when rates look particularly competitive. Thus, the amount they have outstanding can fluctuate between wide margins. The main competition for commercial paper is bills of exchange — acceptances — but these are not available to property companies because bills of exchange are issued to finance trade-backed transactions. Property investment does not count as a trade transaction.

What does a company do if conditions in the commercial paper market are unattractive for further issues at the time it needs to redeem outstanding paper? In the case of smaller or less well known borrowers, investors may expect the company to have some form of committed fall-back facility, which means it would always have access to funds to repay the commercial paper.

This might be a guaranteed line of credit from a bank or could take the form of a multiple option facility (a MOF) with a committed facility. We will be looking at MOFs in the next chapter. Broadly, the committed facility means that a group of banks have arranged to make a specified amount of funds available at a pre-arranged rate over LIBOR if the company cannot raise the finance more cheaply elsewhere. A MOF frequently provides backing for a commercial paper programme.

What about the costs of a sterling commercial paper programme? It might cost £5,000 to £10,000 to set up in the first instance. Earlier, the range of interest rate spreads was comparatively small, from around LIBID for top rank industrial companies to about 12 or 13 basis points over LIBOR — occasionally more — for small or not very familiar companies. Medium-sized property companies probably found themselves near the top end, partly because their cash flow characteristics are not always as attractive as those of industrial companies. By 1990 there had been a general trend for spreads to widen and during the ensuing recession they remained considerably higher than in the boom days of the late 1980s when banks were falling over themselves to lend.

Very large property companies are not necessarily confined to borrowing in the sterling commercial paper market. There is a very much larger and longer-established domestic commercial paper market in the United States for those wanting dollar funds or prepared to borrow dollars and swap into sterling (which can provide cheap finance). Here a rating and a fallback financing facility will be essential. There is also a euro-commercial paper market, which has been tapped by some large property companies.

30 Multiplying the financing options

Newer financing techniques tend to be available first to the very largest corporate borrowers. Then, as they gain acceptance, they gradually become available to borrowers further down the size spectrum.

This is true of the "multiple option facility" or "MOF", which is a flexible method of tapping the money markets for short-term funds. As with commercial paper, the money is borrowed for relatively short periods, but can be rolled over to provide the equivalent of a medium-term loan. As one, say, three-month loan falls due for repayment another three-month loan can be taken out to replace it. It is possible to think of arranging a MOF for sums of around £25m upwards, and large and medium-sized property companies have taken advantage of the technique.

The heyday of the MOF was, however, in the late 1980s when British and overseas banks were vying to lend on very competitive terms. When competition to lend subsided in the subsequent recession, the MOF lost a lot of its appeal to the banks. A resurgence of competition might see a revival.

The MOF should provide cheap funds because it encourages banks to compete to provide the finance. It can take a number of different guises, but in its most common form it consists of two parts: a "guaranteed" or "committed" facility which ensures that the company will get its money when it needs it. And an "uncommitted" facility which may allow the company to borrow on more advantageous terms.

It works like this. Suppose a company wants to arrange a MOF for £100m (this means it can have a maximum of £100m outstanding at any one time) with a life of five years. It instructs an agent — a bank — to set up the facility for it. This arranging bank puts together a syndicate of banks who agree that jointly they will provide up to the £100m required if called on to do so. The interest rate is agreed at this point: not as an absolute amount but as a margin over LIBOR, the normal yardstick of the cost of funds in the money markets.

The agreed margin might be, say, 15 basis points or 0.15% over LIBOR. So if LIBOR is 10% when the company comes to borrow, the finance will cost it a maximum of 10.15% in interest charges, though there are other costs that we will come to later. Borrowing costs will thus rise or fall with movements in LIBOR, but the maximum amount that the company pays in relation to LIBOR stays constant.

We say "maximum amount" because this is not necessarily what the company pays. The second stage of the MOF is for the arranging bank to put together a "tender panel" which will add some additional banks to the original syndicate that agreed the committed facility. When the company wants to borrow, the tender panel banks are invited to bid to provide the finance. Those offering the cheapest rates will make the loans. If the banks in the tender panel do not bid the amount the company requires, or do not bid at a rate below that on the committed facility, the company will need to draw down funds under the committed facility instead. Either way, it is sure of getting its finance.

The advantage of the system is, in theory, the competitive element the tender panel introduces. Banks which are flush with funds at the time can bid to put up the money. Those that are not flush do not need to bid — though tender panel banks that never play an active role will not find themselves popular. So this part of the facility is "uncommitted". The cost of the uncommitted funds will normally be lower than for the committed facility, because the tender panel banks are not entering into an undertaking to supply funds whether it suits them or not.

Sometimes the limit on the uncommitted facility will be the same size as the committed facility. Sometimes the uncommitted facility will be larger. Thus a company might line up a committed facility of £100m and an uncommitted facility of £150m. The maximum it can borrow under this arrangement is £150m, but it is not guaranteed more than £100m.

However, by 1990 it looked as if the tender panel mechanism was becoming less effective as the banks' own return on assets criteria became more aggressive. Borrowers were more frequently thrown back on the committed facility. After the property crisis of 1990-93 and the general recession it took the banks some time to recover from the trauma and it remained to be seen whether competition to lend would revive sufficiently to give the tender panel mechanism a new lease of life.

The "multiple option" in the MOF refers to the fact that the arrangement allows the company to raise its funds in more than one form, and it can pick which suits its convenience or which looks cheapest at the time. As well as straight loans there will normally be bills of exchange — facilities for issuing these are known as "acceptance credits".

A bill of exchange is simply a form of short-term IOU which can be sold at a discount to raise cash, but unfortunately this option is not available to the property investor because, as we have seen, use of bills is limited to trade-backed transactions. There may be facilities for issuing other forms of short-term IOU, and for raising loans in other currencies as well as sterling.

The range of options provided will depend partly on the company's size and sophistication, as well as its geographic spread. One other option is a commercial paper programme, which we looked at in the previous chapter. It may be included in a MOF. But more commonly, perhaps, the commercial paper programme is set up separately but is backed by the committed facility of the MOF. Investors in the company's commercial paper know that, if necessary, it could raise funds from its committed MOF facility to redeem the commercial paper as it falls due.

To calculate the real cost of a MOF you need to allow for all the fees involved as well as the interest rate margin on the funds borrowed. There will be up-front fees for the agent bank and for setting up the committed facility, which can be expressed as an annual rate over the life of the MOF. And there is, of course, an annual underwriting fee for the banks which provide the committed facility. This can be regarded much as an insurance premium — the price of ensuring that funds will be available when required.

The hope is that the tender panel will provide finance at a cheaper rate and that the company will not need to draw heavily on the committed facility. There will therefore be a "utilisation fee" if it does in fact draw more than a certain proportion of the committed facility — perhaps if it draws more than 50%.

This is reasonable from the viewpoint of the lending banks. One circumstance in which they might be called on to put up a fair proportion of the finance is if something happens to knock the company's credit rating after the interest rate margin was agreed at the outset. In these circumstances the tender panel banks may not be keen to bid to provide the funds, or they may only tender at a high rate, so that the burden of providing the finance and taking the risk falls heavily on the underwriting banks.

One further cost reflects the fact that, under prudential rules, banks are required to allocate a certain amount of their own capital against lending commitments. This involves costs for the banks, which are normally passed on to the borrower.

31
Hedging interest rate risks

In a period of volatile interest rates, borrowers increasingly seek protection from the wilder swings. Nowhere is this more true than in the property business, which uses a higher proportion of borrowed money than most businesses, often works on fine income margins and can see its cash flow severely damaged by an unexpectedly large rise in interest rates.

We have already seen that it is possible to fix borrowing costs in advance, relative to one of the common yardsticks for interest rates such as LIBOR. But this is not, of course, the same thing as fixing interest costs in absolute terms. You may know that the most you will pay is 15 basis points over LIBOR. But if LIBOR rises from 8% to 15%, your borrowing cost still almost doubles.

So how can you protect yourself from soaring interest rates? The simplest answer, of course, is to borrow fixed-interest money rather than variable-rate money. Many of our larger companies, including property-based concerns, have been doing this in recent years. With the government a net repayer of debt rather than a borrower in 1989, a shortage of long-dated government bonds began to emerge. This created a fair bit of competition among investors for the those that remained, with the result that prices of government bonds for a time remained relatively high and yields relatively low.

Companies were able to take advantage of this phenomenon by issuing their own long-dated bonds at rates of interest above those on government bonds but still well below the short-term interest rates prevailing at the time. By 1990, long-term borrowing rates had moved right up and for the time being looked less attractive again. In 1993 and early 1994, relatively low long-term interest rates again attracted borrowers, though this time the short-term rates had fallen sharply as well (see graph on page 199). Bond rates subsequently moved up again but fixed-interest borrowers were not always deterred.

Very long-term fixed-interest borrowing may suit property investment companies with revenue-producing properties that they are intending to hold on to indefinitely: a long-term borrowing matches a long-term asset. But it would not be possible for smaller concerns or those with a less established cash flow.

In any case, fixed-interest borrowing involves one significant drawback: you may be locking yourself into a relatively high rate of interest for a very long period. If interest rates fall, you get no benefit from the reduction and could find yourself at a disadvantage to competitors who borrowed at variable or floating rates of interest.

So for some companies the best compromise is to borrow short or medium term at variable rates of interest, but to "hedge" the risk that interest rates will rise further. There are a number of ways this can be done, though always at a price. The important point is to decide at the outset what you are trying to achieve.

Are you trying to fix your exposure to changes in interest rates within very narrow bands? Are you prepared to forgo some of the benefit from a possible fall in interest rates in return for protection against a rise? Or do you simply want a "long stop"; in other words, you are prepared to live with a reasonable rise in borrowing costs but want protection against the possibility of a catastrophic jump? The route you take depends largely on the answer to these questions.

The way you pay for protection varies, too. If, in effect, you are buying an option on interest rates, you pay the cost of the option at the outset. If the option turns out to have been unnecessary, you have forfeited what you paid for it. But it is rather like an insurance premium. It may have been worth the cost to have had the cover. And at least the cost is limited to that up-front premium.

Alternatively, you might enter into an arrangement where what you pay or what you are paid depends on what happens to interest rates in practice. You avoid the up-front cost, but you do not know what the final cost will be.

There are markets in which you can buy interest rate hedging products, and the main one in the UK is LIFFE, the London International Financial Futures and Options Exchange. In practice, relatively few companies make use of it because it may be difficult to find an exact match for the interest rate risk you want to hedge. It is more likely that you will buy an interest rate hedge from a bank, tailored to your particular needs, and the bank itself may then lay off its risk in the financial futures market. These tailored hedges (often known as "over-the-counter" or "OTC" financial products) and similar offerings contributed to the explosion in

"derivative" financial products available from the banking system. By the mid-1990s the banking regulators were worrying about the risks to the banks from this derivatives trade.

Perhaps the most common form of interest rate hedging instrument is the "cap". This means that you put a "cap" or top limit on the amount of interest you might have to pay. In effect it is an interest rate option, and the cost has to be paid at the outset.

Suppose you are borrowing £10m at a variable rate of interest of 50 basis points over LIBOR. And suppose for simplicity that LIBOR is currently 10%. Your money is thus costing you 10.5%. You reckon you could pay up to 12% without suffering too severely. But above that level your business may be in trouble. So you need to set an upper limit of 12% on your interest costs.

Since you are paying 50 basis points over LIBOR, this means you want to protect yourself against any rise in LIBOR above 11.5%. So you buy a cap to protect your £10m borrowing at a "strike rate" of 11.5%. If LIBOR should rise above 11.5% during the life of the cap, the bank from which you bought it reimburses you the difference on a £10m borrowing. So if LIBOR rose to 12% (taking your actual borrowing cost up to 12.5%) the bank would reimburse you the difference between 11.5% and 12% on £10m.

The important point to note is that the cap is separate from the loan itself. You continue to pay your 50 basis points over LIBOR on the loan. The cap simply reimburses you when the interest cost rises above a certain level. And the bank from which you buy the cap does not need to be the same one that made the loan, though it could be. The loan and the cap are two separate transactions.

The cost of the cap obviously varies according to the degree of protection you want (though it can be reduced by selling a "floor" — see below). If LIBOR is currently 10% and you want protection if LIBOR rises above 10.5%, you must obviously expect to pay more than if you are only seeking to protect yourself against a rise in LIBOR above 11.5%. And if you had decided at the outset that you only wanted "long stop" insurance, it would be cheaper still to protect yourself against an increase above say, 13.5% in LIBOR.

As with any other kind of option, the cost of the cap also varies according to how long it has to run. A cap for a long period will be more expensive than one for a short period. Caps can run as long as 10 years, but one or two years would be more common. Though the price of the cap has to be paid at the outset, it can be expressed as the equivalent of an annual rate of interest on the borrowing.

Different influences are at work on short-term and long-term interest rates, represented here by the London interbank rate and by yields on long-dated gilts. Short-term rates, which determine the cost of most bank loans, are frequently manipulated by the government to achieve its economic objectives. Thus rates may be raised to support the currency and squeeze inflation out of the system. When this happened towards the end of the 1980s it helped to provoke the property market collapse of the early 1990s. Long gilt yields, which provide an indication of long-term borrowing costs, respond more to expectations of long-term inflation. Source: *Datastream*

So, in our example, the borrower who pays 50 basis points over LIBOR and bought a cap at a strike price of 11.5% is reimbursed if his own interest costs rise above 12%. But if the cap costs him the equivalent of 25 basis points a year, say, his real maximum borrowing cost including the cost of the cap is 12.25%.

Caps are usually available to cover LIBOR-linked borrowings of £1m and upwards, though for a borrowing linked to base rate it might be possible to cap a much smaller loan. A cap linked to base rate works slightly differently as regards the detail, though the principle is the same.

Finally, if you discover you no longer need the cap while it still has a reasonable time to run, you ought to be able to sell it to recoup part of your original cost. This might happen if you repaid the £10m loan sooner than you expected to and therefore no longer required the interest rate protection.

There is a way of reducing the cost of protection via a cap if you are prepared to forego part of the benefit you might derive if interest rates should, in the event, fall rather than rise. The principle is easiest to understand if you remember that a cap is a form of interest rate option that you buy from a bank in return for an up-front premium payment. You might be able to offset part of this premium cost by selling an option to the bank as well as buying the cap from it.

Take an example. LIBOR is, say, 10% and you want protection on a £10m LIBOR-linked borrowing against the possibility that it might rise above 12%. You feel that interest rates are probably going to move up. So you buy a cap at 12% LIBOR. If LIBOR goes above 12%, the seller of the cap will reimburse you for the difference between 12% interest on £10m and whatever rate LIBOR rises to.

At the same time, you do not think it is very likely that LIBOR will fall below 10% during the period covered by the cap. But should it do so, you are prepared to forego some of the benefit from the reduction in interest charges on your £10m borrowing. You decide, say, that you are prepared to accept that you will get no further benefit if LIBOR should drop below 9%. So you sell the bank a "floor" at 9% LIBOR.

The premium you get from selling the floor can be offset against the cost of buying the cap.

A cap and a floor together are known as a "collar" or a "cylinder". And in our example the collar arrangement means that if LIBOR rises above 12%, the bank recompenses you for any excess. If, however, LIBOR should fall below 9%, under the floor arrangement you must pay the bank the difference between 9% and whatever rate LIBOR falls to. So your interest rate exposure is confined to the range between 12% and 9% LIBOR. Movements outside this range will neither damage you nor benefit you.

Remember that both the cap and the floor are independent of the original £10m borrowing you are seeking to protect and may be arranged with a bank other than the lender, though the bank would be unlikely to be prepared to arrange a floor unless you had bought a cap as well. On your £10m borrowing you pay the market rate of interest. The collar simply reimburses you for the cost above a certain level, or requires you to make a payment to the buyer of the floor below a certain level.

Remember, too, that your actual borrowing cost is almost certainly above LIBOR. If you are paying 50 basis points over LIBOR on the £10m loan and want to limit your actual interest costs to 12%, you would need to arrange a LIBOR cap at 11.5% (or lower if the cost of the cap itself is counted in the equation). Likewise, if you are prepared to forego any benefit from a reduction below 9% in your borrowing costs, you would actually need to sell a LIBOR floor at 8.5%.

It is possible to ask for a "no-cost collar" under which the amount you receive for the floor exactly cancels out what you pay for the cap, though it is unlikely that the terms you would be quoted would exactly match your hedging needs.

The problem with an option, where you pay an up-front premium, is that the finance director may find the idea difficult to sell to his board. The premium is money gone for good, and whether it proves good value or not depends on how interest rates move. Some finance directors claim they will be in less trouble with the board if the company loses out by failing to hedge interest rates than they will be by laying out money on a cap which proves later to have been unnecessary. It is illogical, but sometimes a fact of life.

An alternative approach, which gets round this problem but may bring some problems of its own, is to make use of a forward rate agreement or FRA. This is simply a way of fixing the effective cost of a borrowing (or the income received from a deposit) in the future, though again it works by compensating for the effect of a rise in loan rates rather than freezing the cost of the loan itself.

A company might, say, know that in three months' time it is going to have to borrow £10m for six months. It is worried that interest rates might go up in the interim.

Suppose again that LIBOR is currently 10% and the company wants to protect itself from any increase. It "buys" a FRA covering the appropriate period, though does not put down any cash at this point. FRAs are quoted with a spread between buying and selling price, which covers the cost of setting up the arrangement.

In three months' time, when the company needs to borrow its £10m, LIBOR has risen to 11% and the rate the company will be charged on its loan will therefore be geared to 11%

LIBOR (if it pays 50 basis points over LIBOR the loan will cost 11.5%). But the bank which "sold" it the FRA will compensate the company for the cost of the difference between 10% and 11% LIBOR.

The other side of the coin is that, if interest rates should fall rather than rise, the company's £10m loan will be geared to a LIBOR rate of less than 10%. So if LIBOR is 9% it pays only 9.5% on its £10m borrowing. But in this case it has to pay the difference between 9% and 10% LIBOR to the bank which sold it the FRA. So the FRA means that the company is protected from a rise but gets no benefit from a fall in interest rates.

A company which knows it will have cash to deposit in three months' time and is worried about a fall in interest rates before then would sell an FRA instead of buying one. Again taking 10% LIBOR as the starting point, the company is compensated by the buyer of the FRA for a fall below 10% in LIBOR, but has to compensate the buyer for any rise above 10%.

Buying or selling an FRA is simply taking a bet on the movement in interest rates. If you bet right you are paid, if you bet wrong, you pay out. But failing to hedge a movement in interest rates is probably an even more dangerous bet. The FRA merely offsets the gamble that is inherent in borrowing at variable rates of interest in the first place. If you win on your borrowing because interest rates fall, you lose on the FRA because you have to pay out.

It should be possible to arrange an FRA to cover a period starting as much as 18 months ahead, though shorter periods would be more common.

Before leaving the question of interest rate hedging, there is one more product to cover: the swap. The principle is simple enough, though some of the detail may be complex. Basically, a borrower of floating-rate funds can "swap" its interest payment obligations with a borrower of fixed-rate funds. It therefore locks into a known fixed rate of interest for a given period.

Assume that there are two companies, Superco and Mediumco, each of which needs to borrow, say, £100m for seven years. Superco is the better known concern, is judged the better credit risk and can therefore normally borrow at a lower rate of interest than Mediumco, which is a respectable concern but not a household name.

Superco, say, could normally borrow floating-rate funds at LIBOR plus 0.5% (in other words, at 50 basis points over LIBOR). Fixed-interest money would currently cost it 10.5%.

Mediumco would have to pay LIBOR plus 1% for floating-rate funds and fixed-rate money might cost it 12%.

	Cost of borrowing		*Difference*
	Superco	Mediumco	
Fixed rate %	10.5	12.0	1.5
Floating rate %	LIBOR + 0.5	LIBOR + 1	0.5
Net saving %			**1.0**

Clearly, Superco can borrow more cheaply than Mediumco in all circumstances. But its greatest relative advantage is in fixed-rate money where it has a 1.5% (150 basis point) advantage over Mediumco. For floating-rate money the difference between the two is 0.5% or 50 basis points.

In practice, Superco wants floating-rate funds and Mediumco wants fixed-rate money. Suppose that each raises the kind of finance it wants and suppose also that LIBOR is currently 13%. Superco borrows its floating-rate money at 13.5% at which level (if LIBOR remained unchanged) its annual interest rate cost on the £100m borrowing would be £13.5m. Mediumco borrows its £100m at 12% fixed and its annual interest cost is therefore £12m. The total interest cost on the combined £200m of borrowing is thus £25.5m.

Then suppose, instead, that Superco borrows fixed-interest money and Mediumco borrows floating-rate funds. Superco pays 10.5% for its fixed-interest finance so the annual interest bill on the £100m is £10.5m. Mediumco pays 14% for its floating-rate money which means, if LIBOR remained unchanged, its annual interest bill would be £14m. So this time the total interest cost of the combined £200m borrowing is £24.5m.

If each of the companies borrows in the form in which it has the greatest relative advantage (or the least relative disadvantage) there is therefore a saving overall of £1m in interest costs. This is what provides the rationale for the swap. If the companies then swap interest payment obligations, each ends up the form of finance it wanted, but there is an overall saving which can be shared between them to reduce the cost of finance for each. The stronger party (in this case Superco) will probably take the greater share of this saving.

To see how the savings work out in percentage terms, we need to follow the transaction through. Superco could borrow fixed at 10.5% and Mediumco at 12%, so there is a 1.5% potential saving if the better-rated borrower raises the fixed-interest money. Against this, Mediumco will have to pay 0.5% more for its floating-rate money than Superco would have done. If we knock this off the saving on the fixed-interest borrowing, the net saving on the two transactions is 1%.

Ignore for a moment any costs involved in the swap. Superco raises a £100m fixed-interest eurobond at 10.5%, on which it pays interest in the normal way. But Mediumco, since it wants fixed-interest money, pays fixed-rate interest of, say, 10.7% to Superco (the 10.5% to cover the eurobond interest plus an extra 20 basis points).

Mediumco, meanwhile, has borrowed its £100m at LIBOR plus 1%, and pays the floating-rate interest in the normal way. Superco, which wanted floating-rate money, pays floating-rate interest at LIBOR to Mediumco.

The position of Mediumco is then as follows. It pays LIBOR plus 1% on its borrowing, but receives interest at LIBOR from Superco. Its net interest cost here is thus 1% (100 basis points). The fixed-interest payment to Superco costs it 10.7%. So it has, effectively, the use of fixed-interest money at 11.7% (10.7% plus the 1% net cost on the floating-rate borrowing). It thus saves 30 basis points or £300,000 a year on what it would have paid had it raised fixed-interest money direct.

Superco, on the other hand, pays 10.5% fixed on its eurobond and pays LIBOR to Mediumco. Since it receives 10.7% fixed from Mediumco, it "saves" 20 basis points on the fixed-interest borrowing to offset against its payments of LIBOR, meaning it effectively borrows at 20 basis points below LIBOR. Since, if it had borrowed floating-rate funds direct, it would have paid LIBOR plus 0.5%, its total saving via the swap is 70 basis points.

In practice it is unlikely that the two parties to the swap would deal direct with each other. They would probably each deal with a bank which stood in the middle of the transaction and took a small cut on the way. And they would not necessarily each pay the full interest to the other — a payment for the net difference would pass one way or the other.

The swap does not involve any transfer of capital — it is simply that each party assumes the other's interest payment obligations, with adjustment for the savings to be achieved. So at the end of the term of the swap, there are no capital payments to be made. But it does allow a borrower who can most readily raise floating-rate money to hedge his interest

rate exposure by transmuting it into fixed-rate money at a cost below that of raising fixed-rate money direct.

Nor does a swap necessarily have to be arranged when the original loan is taken out. A company with most of its borrowings at floating rates of interest could decide later that it wanted to reduce its interest rate exposure by negotiating a swap to transmute a proportion into a fixed-rate obligation.

A swap does not necessarily have to run its full course if the circumstances of the borrower change. It can be unwound in various ways to leave the borrower in the same position as if he had not swapped in the first place. But whether the unwinding costs money or actually shows a profit to the borrower will depend on the way interest rates have moved in the interim.

Swaps are available not only for interest rates but for currencies as well. Superco might need dollars, but would have the greatest advantage if it borrowed sterling in Britain where it was best known. Mediumco might be better known as a borrower in the United States, but might require sterling funds. The same principle holds good. Each borrows where it has the greatest relative advantage, then the funds are swapped so that each ends up with the currency it needs. Swaps in which both currencies and interest rate obligations are exchanged are common.

32
Borrowing for the long term

We have looked at short-term borrowing methods for property companies, and the different ways of hedging the interest rate risk that goes with them. At the opposite end of the borrowing spectrum, we now switch to the long-term sources of finance. Take as a starting point the mortgage debenture: the longest-term and most traditional way of raising debt finance via a security issue. It is worth examining in some detail because it conveniently illustrates a number of principles — particularly those relating to security — that crop up in different forms of property company borrowing.

A debenture is in essence a tradeable "IOU" and similar in many ways to the gilt-edged securities (bonds) issued by the government. But the terminology can cause confusion. In America, a debenture normally means an unsecured borrowing. In Britain, it is almost certainly used to mean a borrowing secured on the assets — or some of the assets — of the company. In this way it differs from the otherwise similar unsecured loan stock or the bonds issued in the euromarket (which are usually unsecured).

A mortgage debenture is ideally suited to the needs of many larger property investment companies or to other types of company with a large property base. Brewers are a typical example, because of the large number of pubs they own which provide attractive security for the loan. Typically, a debenture can run for 30 years or more and will pay a fixed rate of interest throughout its life. Companies which own assets they intend to hold as investments for this length of time thus have the opportunity to match the life of their liabilities to that of their assets.

From the investor's viewpoint, a debenture is a long-term asset which can be bought to match long-term liabilities. Thus it appeals particularly to insurance companies. They know what nominal return they will be getting over the life of the debenture and can use it to match known long-term liabilities such as various forms of savings contract.

The bugbear, as with other forms of long-term fixed-interest investment, is inflation. It bites two ways. Investors will be less ready to put up fixed-interest money if they feel that the real value of the loan (and of the interest payments) will be eroded by inflation. Or, at best, they will require very high interest rates to compensate for this risk. The company issuing the loan may be reluctant to commit itself to a high rate of interest for 30 years if it feels that there will be a chance of borrowing at lower rates in the interim.

For this reason the debenture, popular in the 1950s and 1960s, went out of favour in the high inflation era of the 1970s and only began to regain popularity in the latter part of the 1980s when long-term interest rates often looked comparatively low compared to those for short-term money. Long-term rates had climbed back up by early 1990, but by 1993 both long- and short-term interest rates were again significantly lower.

Borrowing via a debenture will often be cheaper than raising cash via an unsecured borrowing, in the domestic market at least, because of the greater security for the lender — lower interest rates normally go with lower risk. It works like this.

The company wants to raise, say, £200m (this is the size, as it happens, of a debenture issue made by Land Securities, the country's largest listed property company, in the autumn of 1988, which serves as a convenient example). The company agrees to charge certain of its properties as security for the loan. The process is much the same as when a homebuyer gives the lender a mortgage on his property as security for the loan.

But there is usually another stage, at least when it is intended that the mortgage debenture will be traded on the stock market. A trustee is appointed who holds the security for the benefit of the investors. This means that the trustee, which will normally be a department of a major insurance company, can act in the interests of all investors in the debenture. It will hold the title deeds to the properties which are charged as security.

If the borrower — the property company — fails to observe the conditions of the loan, and particularly if it defaults on the interest payments, the trustee can negotiate on behalf of investors or — at worst — can sell the properties

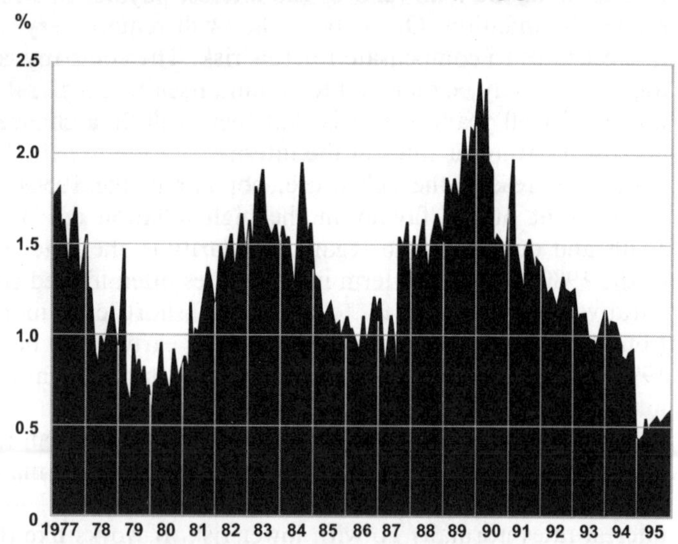

The government – regarded as risk-free – borrows at the lowest rates. Industrial and commercial companies, including property companies, pay more. How much more will depend on the standing of the borrower, but also on the financial climate of the day. The graph shows the margin or spread between long-term borrowing costs for the government, represented by gilt yields, and average long-term borrowing costs for companies, represented by yields on industrial debentures (company bonds). The additional borrowing cost for companies was almost 2.5 percentage points at the turn of the decade when markets were worried about the risk of company failures during the recession. It subsequently dropped very sharply as the economic climate improved. Source: *Datastream*

and repay investors out of the proceeds. If the properties fetch more than the amount of the loan plus interest owing, the surplus goes back to the company. If the properties do not fetch as much as the outstanding amount of the loan, the remainder that is owing is still a debt of the company.

Investors in the debenture are thus secured creditors of the company. And since specific assets are charged as security for the loan, they are fixed-charge creditors. They have the right to be repaid from the proceeds of sale of specific assets, rather than merely having a general priority claim on the assets of the company as a whole. And hereby — thanks to the Insolvency Act 1986 — lies a problem, which we will come to later.

But first, the broad outline. The borrowing company grants a charge over properties to the requisite value. To allow for possible fluctuations in the value of the assets and provide a cushion for the loan, lenders will normally require the borrower to charge assets to a value of at least one-and-two-thirds times the amount of the loan. Thus, in the case of the Land Securities £200m issue, the properties charged were independently valued at £333.45m, or 1.67 times the nominal value of the borrowing. This ratio is known as the capital cover.

Lenders will also, in the case of a property company borrowing, usually want an assurance that the property charged as security produces enough net income at least to cover the interest on the debenture. In the Land Securities case the estimated annual income from the properties was £21.42m, which covered the interest liability 1.07 times. This ratio is known as income cover.

These ratios are fairly typical of what investors will require. But if the property were of lower quality — secondary buildings, for example — investors might insist on both higher income cover and higher capital cover. They will also be wary of buildings which effectively incorporate a trading risk. If you lend against the security of a hotel building you know that the value of that building depends at least partly on the ability of the operator to run the hotel at a profit. The same sort of considerations would apply to purpose-built factories an industrial company used in its manufacturing operations.

The borrower also has to decide, with the help of its financial advisers, what yield it will need to offer to persuade investors to subscribe for the debenture. This will depend on two main factors: the standing of the company combined with the quality of assets providing the security; and, of course, the general level of yields applying to other investments of similar type and similar life.

In practice, the terms will normally be pitched to offer "so many basis points over the yield on the benchmark gilt-edged stock" at the time of issue. A company debenture has to offer a higher yield than a government stock because it is regarded as slightly less safe and it will also be less liquid in the market. In other words, it will not be quite as easy to buy and sell it without moving the price as it would be with a comparable government stock.

33
Yields and redemption yields

We ended the last chapter by looking at the way a property company debenture would be priced in the market relative to a gilt-edged stock. But as the basic mechanisms of fixed-interest yields are often misunderstood, a brief return to first principles may help. Those who are familiar with the concepts will doubtless skip what follows.

The interest rate on a gilt-edged stock (government bond) or company debenture is known as the coupon. If we take a mythical gilt called Funding 10%, the coupon is 10%. In Britain, prices quoted for bonds are for a nominal £100 worth of the bond. If the price is 95 in the market, it means you pay £95 (or thereabouts) to buy a nominal £100 of the stock. The interest rate is expressed relative to the nominal value, not the market price. Funding 10% pays interest of £10 a year for every £100 nominal of the stock, regardless of the price of the stock in the market.

The yield on the stock is not the same thing as the coupon. The yield is the return an investor gets on his money when he buys the stock at the market price.

Suppose that when the government originally launched Treasury 10%, it sold the stock to the public at its par value. In other words, it charged £100 per £100 of nominal value. When the market price is the same as the nominal or par value, the yield to the investor will be the same as the interest rate. If the investor pays £100 to buy the right to receive £10 a year in interest, he gets a 10% return or yield on his money (£10 as a percentage of £100).

For the sake of simplicity we will assume for the moment that Funding 10% is an irredeemable (undated) stock; this means that the government is not under any obligation to repay it at a specific time. What determines the price of Funding 10% in the stock market?

The main factor will be the movement in interest rates generally within the economy. In practice, short-term interest rates (which affect most bank loans) and long-term interest rates (which influence long-term bond yields) are sometimes affected by different factors and do not necessarily move in the same direction at the same time. But, for simplicity, we will assume that short- and long-term interest rates are moving in the same direction. At a time when, say, investors might be able to get a gross return of 9% from a bank deposit they might be happy with 10% on a government stock (which carries the risk of capital loss because its value in the market could fall). If bank deposit interest rose to 11%, they might reckon they needed a 12% return from the government stock. And so on. But if Funding 10% pays £10 a year interest on each £100 nominal throughout its life, how can it offer a return above 10%?

The answer is that its market price has to fall. And this is why rising interest rates generally mean a fall in the value of fixed-interest bonds. The mechanism is a little more complicated than this sounds, because short-term interest rates may not be moving as far or as fast as long-term rates (or *vice-versa*). But if long-term interest rates are moving up, long-term or irredeemable bond prices are probably falling.

Look again at Funding 10%. If the price in the market were to drop to 94, the income yield to a buyer would rise to around 10.6%. He pays £94 to receive £10 a year interest, and £10 as a percentage of £94 is 10.64%. If the price dropped to 83, the income yield would rise to around 12% (the investor pays £83 to receive £10 a year interest and £10 as a percentage of £83 is roughly 12%). And so on.

So if interest rates move up to the point where investors expect a 12% income yield on an undated gilt-edged stock, the price of the mythical Funding 10% will have to drop to around £83 before investors are interested in buying it. Until it gets down to this sort of level, there will be a lot of would-be sellers in the market and no buyers. When more people want to sell than buy, the price goes down. The person who originally subscribed for it at £100 thus has a capital loss, on paper at least.

The yields we have been looking at so far are "income yields", also known as "interest yields" or "running yields". But life is a bit more complex than this, because in practice most fixed-interest stocks and virtually all company debentures will be repaid (redeemed) at some point in the future. The redemption date is given after the name of the security. Thus Exchequer 9% 2002 is a stock with a 9% coupon that will be repaid by the government in the year 2002. Treasury 8% 2002-06 carries an 8% coupon and will be repaid at the earliest in 2002 and at the latest in 2006.

Now suppose that our Funding 10% stock is actually redeemable and will be repaid in the year 2000. So its name is Funding 10% 2000. Suppose, too, that you could buy it in the market at 90, at which price the interest yield will be 11.1% (£10 interest as a percentage of £90).

Suppose that we were looking at the stock in 1990 when it was due to be repaid in 10 years. And it would be repaid at its nominal value of £100. So anybody who bought it in 1990 and held on to it until it was repaid at £100 in 2000 would have a capital profit of £10 on his £90 outlay, as well as the interest yield he received each year in the interim. In other words, his total return would not have been limited to the interest income he received. There was also the "gain to redemption" to consider. The combination of interest yield and this notional gain to redemption is known as the "redemption yield". It is the total return to the investor if he buys the stock at £90 and holds it until it is redeemed.

If the £10 profit were notionally allocated over the 10 years that had to run until the stock was redeemed, it is worth £1 a year on top of the £10 of interest the investor collects. In practice the sums are rather more complex than this and require compound interest calculations. But the £10 gain to redemption means that the actual redemption yield is around 11.75% against the running yield of 11.1%. So 11.75% is the total return to the investor buying the bond at a price of 90 in the year 1990.

This redemption yield rather than the interest yield alone is the relevant one in deciding the terms of a bond. So if it is decided, say, that a property company debenture needs to yield one percentage point or 100 basis points over a gilt-edged stock that matches its features fairly closely, it is the redemption yield on the gilt-edged stock that is relevant. Normally, there will be one particular long-dated gilt-edged stock, known as the "benchmark gilt", which is taken as the yardstick for long-dated debentures. So if the benchmark gilt offers a redemption yield of 10% on the day that the terms of the debenture are fixed, the debenture needs to provide a redemption yield of 11%. This is because the issuer or his adviser has decided that, to attract investors, the bond needs to offer a margin of 100 basis points or one percentage point over the gilt yield.

In this case the calculation is easy. The debenture could carry an 11% coupon and be priced at par (£100 for £100 nominal of the stock). But in practice the redemption yield on the benchmark gilt is unlikely to be a convenient round number on the relevant day. Suppose it is 10.15% so that the debenture has to offer a redemption yield of 11.15%. In this case the debenture might still carry an 11% coupon, but would be offered for sale at a price slightly below its nominal value ("at a discount"). This would mean that the redemption yield was higher than the coupon. The precise issue price would be calculated to provide an 11.15% redemption yield.

The Land Securities debenture, launched towards the end of 1988, that we looked at in the previous chapter carried a 10% coupon, but was sold at a price of £97.336 per £100 nominal. Since it was redeemable in the year 2030, this offered a redemption yield of 10.27%, which was 60 basis points over the redemption yield on the benchmark gilt-edged stock on the day the terms were finalised.

34
Problems of security

Having looked at the basic structure of a mortgage debenture issue and at the market mechanisms for pricing it, we need to examine a few of the details.

The attraction — for the borrowing company — of a bond secured on specific properties is that it does not restrict the company as a whole in its financing policies. Because of the security offered, it should also be cheaper than unsecured debt, at least in the domestic market. Where debt is not secured on the company's assets, investors will probably insist on restrictions on the company's overall level of borrowings.

The disadvantage of secured borrowing is that the company loses some flexibility in its handling of the properties that are charged as security for the loan. Clearly, you cannot remove the security by selling the properties.

But some flexibility will normally be built in. There is probably a clause that allows the borrowing company to remove properties from the pool of charged properties, provided it substitutes others which are acceptable to the trustee and provided it maintains income and capital cover for the debenture.

If, over the years, the value of the properties in the pool rises substantially, the position arises where capital cover for the debt far exceeds the stipulated minimum (one-and-two-thirds times, in our example). In this case the borrower is probably allowed to remove some of the properties from charge, provided the capital and income cover requirements are still met. Alternatively, the borrower can probably use the excess security as backing for further issues of debenture stock identical to the original issue and ranking alongside it ("fungible" with it), provided there is sufficient capital and income cover for all the stock in issue.

The other side of the coin is that, if the value of the property charged as security were to fall significantly, the trustee could insist that the company topped it up by putting further properties into the pool. The trustee can call for valuations to establish the value of the properties that are charged. This happened quite frequently when property values collapsed in the 1990-93 period, though the issuers of the debentures did not generally go out of their way to publicise the fact.

Though a debenture still offers excellent security for the lender, the 1986 Insolvency Act has introduced a complication whose ramifications have taken some time to sink in. Since it affects many other forms of secured borrowing as well, we need to examine it briefly.

Prior to the 1986 Act, if a borrower defaulted on the terms of debenture with a fixed charge over specific properties, the trustee for the debenture could simply sell the properties to recover the loan. Since the new Act came into force, this is not always possible.

The problem is that the 1986 Act gives creditors with a floating charge different rights from creditors with a fixed charge. A floating charge is a charge over the assets of the company as a whole (including current assets) rather than a charge on specific assets.

If a company gets into financial difficulties, but thinks it might be able to resolve these (or secure a better outcome) if it could prevent its creditors from closing in, it can apply to have an administrator appointed. If its application is granted, the administrator will present a plan for dealing with the situation, which might involve continuing to run the company as a going concern and perhaps only selling off assets over a period of time. If this happens, it effectively freezes the creditor with a fixed charge. He loses his ability to move in and sell the relevant assets so, though the security for his loan still exists, control over that security — and decisions on when to sell it — pass to the administrator.

The creditor with a floating charge, on the other hand, has the right to appoint an administrative receiver and to block the appointment of an administrator by the company itself.

This system has far-reaching implications. It means that lenders will try where possible to obtain a floating charge as well as a fixed charge, so that they can avoid the risk of being frozen. But from the borrowing company's point of view, an additional floating charge may remove some of the attractions of the fixed-charge borrowing. No longer is the charge limited to a designated pool of assets. There are various ways of attempting to deal with the problem. Perhaps the borrower could grant a floating charge over the assets of a specific subsidiary that owned the properties subject to the fixed charge, rather than charging the assets of the group as a whole.

In the case of the Land Securities 10% first mortgage debenture, issued towards the end of 1988 which we have taken as an example, there was a fixed charge but no floating charge. And the company was therefore obliged to point out that "The Trustee's powers of enforcing the security will in certain circumstances be limited by the Insolvency Act 1986". Because of the restrictions on enforcing the security, investors may demand a somewhat higher yield on a debenture that does not offer a floating charge as well as a fixed one.

Most large debenture issues by UK companies will be listed on the Stock Exchange and can be bought and sold in much the same way as the government's gilt-edged stocks. Some smaller issues might simply be placed with (sold to) a limited range of investors without the intention of providing for a market in the stock.

Alternatively, the company could give existing investors the first chance of subscribing for the stock, but since it is a debt security rather than equity or a convertible, it is under no obligation to do so. The Land Securities issue, though listed on the Stock Exchange, was initially simply placed with investors. As is the case with issues in the domestic market, it was available in registered form. In other words, owners of the stock are registered with the company, which therefore knows who holds it. Issues in the euromarket, which we will come to later, are more likely to be made in bearer form. In this case, the stock certificate itself is proof of ownership.

However, by the mid-1990s the distinction between sterling-denominated bonds issued in the domestic market and in the euromarket was breaking down. In the second half of 1995, property company British Land made an issue of a forty-year first mortgage debenture bonds, issued both through euromarket and domestic market mechanisms — investors could take their choice between the (bearer) euromarket bond or the (registered) domestic bond.

One final technicality. Like some of the government's own bonds, a debenture issue may initially be offered in part-paid form. This means that the investor does not have to put up the whole of the price immediately. He pays a first instalment, then puts up the remainder of the money in one or more further instalments some months later.

In the case of the Land Securities issue, for every £100 nominal of stock the investor was required to put up £30 initially. The remainder of the £97.336 purchase price was payable a few months later.

This is a feature which sometimes adds to the initial appeal of a stock by introducing a slight speculative element. Suppose interest rates fell after the initial payment was made but before the second. And suppose they fell to the extent that the price of the stock might be expected to rise to, say, £101 in the market. This would be a premium of £3.664 over the issue price. So, in its £30-paid form, the price of the stock might be expected to rise to around £33.664. This is a profit of over 10%, so while a stock is only part-paid a comparatively small movement in interest rates can produce quite a significant swing in value one way or the other. A part-paid stock is thus a highly-geared investment.

35
Tapping the euromarkets

The property company debenture we looked at in Chapters 32-34 was a secured borrowing raised in the domestic market. But larger companies also have the opportunity to raise long-term unsecured loans, in either the domestic or international market.

The international market, or euromarket, needs a brief introduction. In theory it is a market in "stateless" money — money that is held outside its country of origin. Thus deposits of US dollars in a German bank would be "eurodollars", pounds sterling held in Holland would be "eurosterling", and so on.

These stateless funds can be borrowed and lent much like domestic funds. But the market in eurocurrencies is not subject to the controls of any one domestic financial authority. London is still the main centre for eurocurrency dealing, but British banks are not the main players in the market, which is dominated by American and Japanese banks. There is no central marketplace — deals are arranged over the telephone — and funds can be borrowed in most of the major world currencies.

The eurocurrency markets provide a parallel for most of the products and facilities available in the domestic financial market. Thus money can be borrowed very short term in the form of euro-commercial paper or very long term in the form of a 20-year eurobond. And it can be borrowed in the form of bank loans as well as security issues.

But there are some fundamental differences between the euromarket and the domestic market. First, "euro" borrowing is generally unsecured. Without the security, the reputation and standing of the borrower is all-important. Thus, major internationally known corporations have no trouble in tapping the market, but it is not so accessible to smaller companies known only in their domestic markets.

Domestically, property investment companies have always been regarded as exceptionally secure because of the strength of their assets. But in the euromarkets, where investors may be looking more at cash flow and are less familiar with the structure of the British property company, they may not always be rated so highly. So it will be mainly the very large property companies that tap the euromarkets.

However, with changes in the rules for issue of domestic sterling bonds, the distinctions between domestic and "euro" markets have largely broken down for sterling issues. The "euromarket" stands more for an issuing and trading techni-

que than for a totally separate market, and most British banks and securities houses will deal almost interchangeably in domestic and euro issues of sterling bonds. We saw in the previous chapter that a bond may nowadays be issued in both "euro" and "domestic" versions, as with the 1995 British Land issue. Eurobonds are normally listed on a stock exchange — probably London or the lightly-regulated Luxembourg market — but most of the dealing is direct between traders rather than through the exchange mechanisms.

A £100m issue by Hammerson, one of Britain's major property development and investment companies, early in 1989 provides a convenient example of the property company eurobond issue. The bonds, issued at £99.888 for £100 nominal, carry a 10.75% coupon and are repayable in the year 2013. As with most eurobonds, the denominations are fairly large — £10,000 and £100,000 in this case — and the bonds are issued in bearer form. Thus the owners do not have to register with the company. UK income tax on interest payments is not deducted at source by the company. This has an appeal for investors who do not wish to share too much information with the tax man.

In the absence of specific security for the loan, there are some constraints imposed on the company's behaviour. So while the company does not tie down specific properties, it might slightly restrict its financing options in the future. In the Hammerson case, the company agrees that its net borrowings (total borrowings less cash and investments) will not exceed 1.5 times adjusted capital and reserves. And it agrees to limit its secured borrowings to 0.5 times capital and reserves.

Hammerson also undertakes that it will not dispose of assets representing more than 30% of the group's total assets within a specific period — the so-called "Tickler" clause. If it breaches any of these "covenants", fails to meet interest payments or has a receiver appointed to any substantial part of the business, the bonds will become repayable at par together with accrued interest.

"Restrictive covenants" of this kind have become more important to investors following the trend towards "leveraged" buyouts or takeovers, in which a company is bought largely with borrowed money. Clearly, if massive additional borrowings are added on, it may degrade the standing of the company's existing bonds — a process known as "event risk". But it is doubtful if, in practice, the covenants always provide the degree of protection required.

In the case of the Hammerson eurobond, the company itself also has the option, in some circumstances, to redeem the bonds before they become due in the year 2013. If the tax regime changed and it found it had to provide for withholding tax on interest payments it would have the right to redeem at par.

And if, for whatever reason, the company simply wanted to withdraw the issue, it would have the power to do so, but only on fairly penal terms. This would be likely to crop up only if the company wanted totally to restructure itself or its financing, and considered it was worth paying the penalty to have the freedom to do so. The formula for working out the redemption price in these circumstances is related to gilt-edged yields at the time and is a little complex. But in effect it would mean that the issue would have to be repaid at a significant premium over its market value at the time.

Some investors are a little cynical about the protection afforded by this kind of clause, pointing out that if at any time a company wanted to redeem its bonds, it could do so by engineering a technical breach of one of the other conditions, in which case the loan would become repayable only at its par value. Clearly a company such as Hammerson, or one which wants to tap the bond markets in the future, would not resort to this kind of manoeuvre. But if a company falls prey to a takeover predator there is less confidence as to how the new owner might behave.

Other conditions in the Hammerson issue are not dissimilar to those for a secured domestic issue. There is a trustee to look after the interests of bondholders. And the company can make further bond issues provided it does not breach any of its borrowing limits. The bonds are listed on the London Stock Exchange though, as we have seen, dealing in them is likely to take place outside the exchange's mechanisms.

Companies are not, of course, restricted to borrowing in sterling in the euromarkets. They may borrow in other eurocurrencies such as, say, US dollars or German marks, either because they need the currency in question for their overseas operations or because they can reduce their borrowing costs by raising the money in one currency and swapping it into another.

Nor are they restricted to borrowing fixed-interest money, or borrowing for very long periods. Bonds may be issued with a life of just a few years. Or, instead of a fixed-interest bond, the company might issue a floating rate note (FRN). In this case interest payments change with the movement in a benchmark interest rate, usually LIBOR.

There are numerous other permutations on the eurobond theme, most of which have their counterparts in the domestic bond markets. Potentially the most important for the property company is probably the euroconvertible bond, which we mentioned in Chapter 20 and will examine more closely in the next chapter.

36
Warrants, options and complex convertibles

There is no limit to the complex features that can be built into a financing package, and new twists are emerging all the time. But if you once grasp the principles, most of the complexities fall into place.

Various forms of option and warrant are the building blocks of many of these packages, and it is these we need to look at first. Start with an option on a share. A "call" option gives its owner the right to buy a share in the future at a price fixed today, and it is most easily illustrated with an example.

Suppose the shares of Payola Properties are currently quoted at 200p in the stock market. You think there is a very good chance that they might rise to 260p within the next six months. Instead of buying the shares themselves, which would require quite a bit of cash, you might decide to buy an option giving you the right to buy one Payola share at any time within the next six months at, say, 220p (the exercise price or striking price).

Suppose you pay 20p for the option. Currently this option has no intrinsic value. If the Payola price rises to 230p, there is a value in the right to buy for 220p a share you could immediately sell for 230p, but you are still out of pocket because you paid 20p for the option in the first place. If the Payola price rises to 240p, you are breaking even. If it rises to 260p, you can sell for 260p a share which only costs you 220p, so you make a profit of 40p.

From this you have to knock off the 20p your option originally cost, but (ignoring transaction costs) you are still showing a profit of 20p on an outlay of 20p: a 100% return on your money for a rise of only 30% in the share price. Of course, if the Payola price fell instead of rising, you would have lost the whole 20p you laid out on the option.

The owner of Payola shares who sold you the option was probably somebody who took a different view from you of the prospects for Payola, or would have been quite happy to sell the shares at 240p anyway (the 220p exercise price plus the 20p premium he received from selling the option). We have taken a "call" option because it is the easiest to illustrate but, if you thought the market was going to fall rather than rise, you could equally well have bought a "put option" which gives you the right to sell a share at a predetermined price.

In Britain, the options themselves can be bought and sold much like shares in the Traded Options Market, which is now part of the Liffe financial futures market. The price of a call option will tend to rise if the value of the underlying share goes up. The price of a put option will rise as the share price falls.

The point to note is that options, as described here, are not created by the company. If you buy an option on Payola shares, the money does not go to Payola. It goes to the owner of Payola shares who granted you the option.

But there is another instrument called a warrant which is created by the company and brings in money for the company itself. Payola might decide that, rather than issuing new shares for cash now, it will instead sell warrants giving the right to subscribe for new shares at a fixed price in the future.

If the terms of the warrant were similar to the option we have examined, Payola would get 20p now for each warrant it created and sold and would get a further 220p for the share for which the warrant owner eventually subscribed (always assuming Payola's share price rose above 220p and made subscription worthwhile). Meantime, the warrants would probably be traded in the stock market just like shares, and their price would rise and fall to reflect movements in the Payola share price.

Though the description we have given is the traditional one, sometimes nowadays the terms option and warrant are used almost interchangeably. But a warrant created by a company will normally have a much longer life than an option, probably running for a number of years, whereas a traded option (in Britain) has a maximum life of nine months.

The point about warrants is that they are often not issued for cash, but as a "sweetener" to some other kind of issue. Payola might want to issue a fixed-interest bond, but reckon the bond would be more attractive (or the interest rate would be lower) if it gave away, say, one warrant to subscribe for a share with every £10 worth of the bond. When this happens, the bond and the warrant are probably subsequently traded separately.

Nor are warrants always restricted to giving a right to subscribe for shares. The warrant might give the right to subscribe for another bond at some point in the future, on terms fixed today. In this case the warrant offers a bet on interest rates rather than on share prices. These techniques are quite common, particularly in the euromarkets.

Now look again at a convertible loan stock or convertible bond (it is more likely to be called a bond in the euromarket). Say that Payola's share price is 175p and it issues a convertible with a 7.5% coupon. Each £2 nominal of the stock converts into one Payola share, so the conversion price is 200p. The "conversion premium" (difference between the share price and conversion price at the time the issue is announced) is just over 14%.

Effectively, an investor who buys the convertible stock is buying two things: first, a bond that pays 7.5% a year; and, second, an option to subscribe for Payola shares at 200p. But instead of putting up 200p in cash, he will surrender £2 nominal of the bond. The value of the convertible is a mixture of the bond and option elements.

In the Payola case the bond will almost certainly be turned into shares eventually — Payola's share price has only to grow 14% or so and has a number of years to do it.

But a form of convertible quite popular in the euromarket, known as the "premium put convertible" poses problems. The coupon on this convertible bond is, say, 6%. Suppose again that the issuing company's share price is now 175p and the conversion price is 200p.

However, to appeal to euromarket investors more concerned with bonds than shares, the issuing company guarantees that they will get a minimum total return from the time of issue of, say, slightly over 10% a year. By doing this, it gets away with a lower basic coupon on the convertible bond. If all goes well, the share price will rise and investors' gain on conversion, combined with the 6% a year interest, will provide at least the 10% annual return.

But as a fallback, if the share price does not rise fast enough, investors can sell the bond back to the company at a premium (this is a put option for the investor) which will give them their overall return from issue of 10%-plus. Say the company agrees to buy it back at £125 after five years. In this case, at the five-year point, the investor calculates that he can convert £2 nominal of the bond for one share as before. But since he can sell £2 nominal of the bond back to the company for £2.50, the share price has to be over 250p to make conversion a better bet than selling the bond back. In other words, the effective conversion price is rising each year and this moving target means that the share price has to rise much further to prompt conversion, and conversion therefore becomes less likely.

Since the issuing company is normally banking on the fact that the convertible bond will be turned into shares (and thus become permanent capital) at the end of the day, this can throw the calculations badly out if the share price is sluggish. So a whole range of extra features may be built into the terms of the stock to make conversion more likely. "Rolling puts" give the investor several chances to sell the bond back to the company at rising prices, so he feels less need to grab the first opportunity. The company may take the power to freeze the effective conversion price after five years to encourage conversion, and pay extra annual interest on top of the 6% coupon instead, and so on.

The basic problem, as with so many complex financing methods, is that they attempt to reconcile the apparently irreconcilable needs of different classes of borrower and investor. The company issues the premium put convertible bond, looking on it as deferred share capital. The investor may well look on it simply as a bond with a guaranteed minimum 10% return and a possible equity option on top. These and other complexities are not unfamiliar in the euromarkets. Inventing weird and wonderful financing products generates fees for bankers.

37 Project loans for development

The debentures, loan stocks and eurobonds we have been looking at are all "securities" — forms of IOU note issued by property companies and others which can subsequently be bought and sold among investors. But this route is generally not open to smaller companies and the bulk of lending for property development — even to the larger companies — is in the form of straight loans rather than securities. You borrow from a particular bank or group of banks and pay the money back to them at the end of the day.

Much of this lending is against a particular property project, rather than "corporate" lending to the company as a whole. Property development and trading companies, we have already seen, will tend to set up a separate company for each major development and in the past would often structure it in such a way that it did not technically qualify as a subsidiary.

The borrowings of these "off balance sheet" companies did not therefore have to be shown in the group balance sheet. This practice will probably continue, at least in the case of a genuine 50-50 joint venture, even though the definitions of a subsidiary company have now been tightened in Britain to bring some other off balance sheet vehicles into the group accounts.

If a lender is advancing money against a particular development, his considerations are rather different from those of a lender who is advancing money to the company as a whole. The corporate lender is obviously concerned with the strength of the company: its assets, profits and cash flow.

A bank which lends against a particular project, however, looks mainly to that development to provide the security for his loan. So he needs a great deal of detail on the development: estimated costs, rents, completed value, time required for development, and so on. He also needs to have confidence in the ability of the developer to complete the project satisfactorily and let it.

If the development provides the only security for the loan, remember that we would talk about "non-recourse" finance. Should the development go wrong, the lender has no recourse to the parent company or other parties for his interest or repayment of his loan. In practice, "limited-recourse" lending is more common. We will look at what that implies in a moment, and it is worth noting that when commentators refer to "non-recourse" finance, in practice they often mean "limited recourse". If, on the other hand, the parent company gives full guarantees for the borrowings undertaken by the development company, the lender then has "full recourse" to the parent and his loan is more in the nature of corporate than project finance.

While limited-recourse lending is usually for development projects, "holding" or "investment" loans are now becoming common as well. These provide finance for a completed and revenue-producing property, allowing the owner to hold on to it as an investment rather than being forced to sell it to repay the development loans.

Sometimes the project loan for a development may contain a clause allowing it to be transmuted into an investment loan when the project is completed and let. This investment loan probably then runs at least up to the time of the first rent review on the property after five years. Depending on the scale of the development project, the development and investment loans together might thus run for seven to nine years.

Before we look in detail at the structure of a development loan, there are a few other points to note. The money may be borrowed from a single bank. But larger loans — £10m or £20m upwards — are more likely to be syndicated among a group of banks, each of which therefore bears only a proportion of the risk of the project. And the loan might consist purely of senior debt, or it might also incorporate a layer of "mezzanine" finance.

To see where mezzanine debt fits into the picture, it is easiest if we look at the financing structure of the project as a whole. Bank lenders would normally require the developer to put in some "equity" money of his own as a cushion for their loans — if the project hits problems this equity contribution is lost before their loans are at risk. In the case of a project being undertaken by an off balance sheet subsidiary, this equity contribution probably comes from the parent company.

There are varying views on how much of the cost of a project lenders will be prepared to advance as "senior debt"

(we are talking here of a good quality speculative development). The size of the senior debt can also be expressed in two ways: as a proportion of the cost of the project or as a proportion of its estimated value when completed and let. Since finance for speculative development virtually disappeared during the property market recession that began in 1990, we will look back at the pre-recession pattern.

Take the cost yardstick first. It depends on whom you talk to, but in the development boom of the late 1980s the senior lenders might have been prepared to put up between 75% and 85% of the cost of a development. If they were looking at the estimated end value, they would have been unlikely to want to provide loans equal to more than 67% to 70% of value at their own risk. Generally, their loan would have been determined by whichever calculation gave the lower figure.

For simplicity, we will take as an example a development costing £10m (in practice the figures may be very much larger). Assume that banks are prepared to provide 80% of cost, or £8m, as senior debt. On the face of it, this means the developer has to find 20% of the cost or £2m from his own resources — his equity contribution.

However, if the developer wants to borrow more than 80% of the project's cost, he may be able to do so. But lenders would want a higher return on the portion over 80% to compensate them for taking part of the risk that would normally be covered by equity capital. The compensation might simply be in the form of a higher interest rate on this slice, but more probably it would include a share in the profits of the development. This would be mezzanine finance — so called because it is an intermediate level between debt and equity.

The profit share could be arranged in a number of ways (we look more closely at one form of profit-sharing arrangement in the next chapter). The lenders might be granted an equity stake in the project; or they might be granted options on the equity; or they might simply get a share of the surplus over cost that the project throws up at the end of the day.

This brings us back to the question of "limited recourse". With a non-recourse loan, the security for the lender is:

- The development itself.
- The cushion of equity provided at the outset by the developer.

In practice, the banks normally want rather more. Typically, they might require a guarantee from the parent company that the development will be completed — they would not want the responsibility of sorting it out if the development company walked away from the project. And they probably also want limited guarantees for interest overruns and cost overruns. If the project takes some time to let and produce revenue after it is completed, they want the parent to agree to cover the interest for a limited period. This would be a limited-recourse loan — the parent does not guarantee repayment of the loan itself.

Most bankers with a specialist property lending arm will have some property expertise in-house but, certainly for the larger loans, they will require a valuation of the project from an outside firm of surveyors. This is normally in addition to the developer's own appraisal of the project, and the cost obviously has to be met by the borrower. The project also has to be monitored throughout its duration.

Development loans will normally be at a variable rate of interest, probably expressed as a margin over LIBOR. The cost again depends on whom you talk to, but for a limited-recourse loan for a large office development, in late 1989 the margin on the senior debt would probably have been at least 100 basis points (one percentage point) over LIBOR and could be several times this level for smaller developments or other types of project.

In addition to the basic interest, the developer has other fees to meet. There will be a one-off fee for setting up the loan; a "non-utilisation fee" or "commitment fee" on the part of the loan that has not yet been drawn down; professional fees incurred by the banks; and the fees for the monitoring surveyor. In the case of mezzanine finance or other profit-sharing arrangements there is also the cost of the profit share to take into account.

On the other hand, with some development loans the interest rate margin may reduce if the developer prelets the building during construction, and reduce further if he also presells it ahead of completion. This is because both events substantially reduce the banks' risk.

38
Loans with profit share

We have already looked at various ways of sharing profits in a property venture between entrepreneur and financier. The aim in most cases is to reduce the cost of debt finance, or reduce the equity contribution required, in return for conceding a portion of the profit from the project. Use of convertible stocks, leasebacks, complex lease structures to apportion the rental flow — all are profit-sharing techniques of one kind or another. But straight bank loans may also be structured to give the lender a share in ultimate profits.

However, before examining the details of a profit-sharing agreement, a distinction needs to be made. Profit-sharing can be structured so that the two (or more) parties share both the risks and the rewards of the operation — this might be known as a "side-by-side" arrangement. Or it may be structured so that the entrepreneur takes all of the equity risk and the financier simply takes a share of the rewards if the project is successful. Profit-sharing arrangements with bankers fall largely into this second category.

The principle is easiest to illustrate with an example. Take again a development estimated to cost £10m and assume that bankers are prepared to provide £8m, or 80% of the cost, in the form of senior debt. The developer thus needs to find the remaining £2m.

If he is prepared to pay the premium for "top slice" mortgage indemnity insurance (see Chapters 39 and 41), he might persuade the original lenders to put up as much as a further £1.5m as senior debt. Alternatively, without mortgage indemnity insurance, he might raise £1.5m as mezzanine finance which needs to offer a considerably higher return to compensate for the higher risk. In either case, the amount of equity he needs to put in himself is reduced to £500,000.

If mezzanine finance is used, it can offer its higher return in one of two ways. It may simply pay a considerably higher rate of interest than the senior debt — if the rate on the senior loans is 100 basis points over LIBOR, the mezzanine might need to offer 500 basis points over LIBOR. Or it may be entitled to a rate of interest plus a share in the profits.

How big will the profit share be? This depends on two main factors: the return — "internal rate of return" or "IRR" — that the lender requires on his funds, and the expected profitability of the project. If the project is likely to be an exceptionally profitable one, the lender may require only a comparatively small percentage of the ultimate profit to give him his target rate of return. With a more marginal project he may require a considerably higher proportion — perhaps as much as 50%. The IRR he requires on his money will depend on the proportion of project cost he puts up above the 80% provided by senior debt. And it will depend on the business climate.

Early in 1990, financiers of the top slice were looking for a target rate of return in the 35% to 40% region. A year earlier, when the property market in Britain was still booming and competition to lend was strong, target rates of return would have been considerably lower. By 1995 the desired rate was somewhat academic, as so little speculative development was being financed. But doubtless we will eventually swing full circle.

By and large, the big commercial banks which provide the senior debt for a development are not very interested in profit-sharing arangements. Provided they stick within their "80% of cost" criterion, the loan can be fairly easily evaluated as a banking risk and it does not require a massive amount of in-house property expertise. It would also be very difficult to syndicate on conventional lines a loan including a profit-share element.

Profit-sharing arrangements are mainly the province of more specialist lenders with in-house property expertise and experience. Some may provide the whole of the finance for the project: senior debt and mezzanine. But if the project is large and they want subsequently to syndicate the senior debt, it will need to be clearly distinguished from the mezzanine with its profit-share arrangements.

Other specialist property finance operators provide only the mezzanine/profit-sharing slice, and leave the senior debt to others. And there are still further specialist property finance organisations which do not provide funds themselves but structure a scheme for a developer and line up the finance in return for a fee.

We took the example where senior debt provides 80% of cost, mezzanine debt the next 15% and the developer puts in 5% as equity. There are plenty of stories of developments where the developer puts in no finance of his own at all, but these would not appeal to the more prudent lenders. They will insist on some contribution from the developer, if only to ensure that he suffers some pain himself if the project goes wrong. It makes it less easy for him to walk away from a sour development.

For several reasons, a profit-sharing arrangement needs to be carefully structured. The British tax regime means that you need to be very careful not to create debt where the rate of interest is geared to the profits of the enterprise — it would be treated for tax purposes in much the same way as equity and the tax relief on the interest payments would be lost. Additionally, there might be some question as to the

status — from the security viewpoint — of a mortgage paying profit-linked interest (but see "convertible mortgages" in glossary).

So a common way of arranging for the lender to receive his profit share is in the form of a fee. The loan documentation would provide for the payment of a fee related to the ultimate profit.

Next, the "profit" to which the fee relates needs to be very tightly defined at the outset. One way to avoid arguments on this score is to provide for the lender to receive a percentage of the amount by which the proceeds of selling the completed development exceed the originally budgeted costs. This has the added advantage — from the lender's viewpoint — that the developer suffers if he allows costs to overrun, but benefits if he can keep costs below the original estimate.

Finally, the profit share will not necessarily be a straight percentage split. It might be structured so that, say, the developer receives the first 10% profit on cost, the lender receives the next 5%, and above that level profits are split equally between the two or in some other proportion.

For our profit-share example we have taken a development. But it could equally well be a refurbishment or even a straight investment property purchase. However, most of the specialist lenders who require a profit share reckon to turn their funds over fairly regularly — typically, they might expect to see their money back within two or three years.

This poses a problem if the developer wants to hold on to his property on completion, or if the profit-sharing loan is, say, for the speculative purchase of a reversionary property where the owner might want to hang on after the rent review. The normal course in these circumstances would be for the owner to refinance the mezzanine debt on a longer-term basis after the time-span originally envisaged, in which case the mezzanine lender's profit share would be calculated by reference to a valuation rather than sales proceeds.

For a developer, conceding a profit share to the lender has the obvious disadvantage that he does not enjoy all the fruits of his project. On the other hand, it may allow him to undertake a greater volume of development with the (usually limited) equity capital at his disposal. Keeping, say, 65% of the profit on two developments may be better than having 100% of one.

39
Lenders and loan enhancements

In the aftermath of the 1990-93 property crash, finance for speculative development became virtually unavailable. Even by the latter part of 1995, development finance — if available at all — would normally be provided only against pre-let projects. The pattern of lending we describe here is therefore the pattern that applies in more "normal" times, and applied particularly in the late 1980s. Whether the pattern will resume in its previous form when widescale development activity finally resurfaces after the slump remains to be seen.

In "normal" times, property lenders for major development projects fall into two main categories. There are those who provide mainly the "senior debt" — say, the first 75% to 85% of the cost of the project. These would include major British clearing banks and many of their overseas counterparts. They are the banks with a large deposit base and plentiful funds to lend.

Then there are the merchant banks and specialist property lenders who might be more inclined to take an element of equity risk by putting up mezzanine finance or entering into profit-sharing arrangements. Their total funds are probably rather smaller.

In practice the two categories may overlap. The same institution might be prepared to put up or arrange the senior debt and also provide a more risky mezzanine level. Merchant banks (similarly to the clearing banks and major foreign banks) are able to put together a syndicate of senior lenders, though they will not necessarily put up much or any of the senior debt themselves.

If a bank agrees to arrange syndicated finance of, say, £50m for a development project, there are two ways the deal could be arranged. The bank could approach the syndication on a "best endeavours" basis, trying to put together a syndicate of banks, each of which would agree to put up a specific proportion of the money required. Or it could initially take the whole risk itself by contracting to provide the developer with the whole £50m at an agreed margin over LIBOR, only later "selling down" the loan by getting a syndicate of other banks to take a portion each. Clearly, the second procedure is faster for the developer but carries more risk for the bank which is temporarily underwriting (agreeing to provide) the whole loan.

Protection against loan losses. Where a developer wants to borrow more of the cost of the project that the 75% to 85% that bankers would normally be prepared to advance as senior debt (the proportion could anyway be higher if the project was pre-let or pre-sold), we have seen that he can raise additional money as mezzanine debt, probably by conceding part of the profit from the project. Another approach that was popular in the 1980s was to take out insurance on the project.

The process is much the same as with a mortgage on a house. If you want to borrow more than 75% of value, you

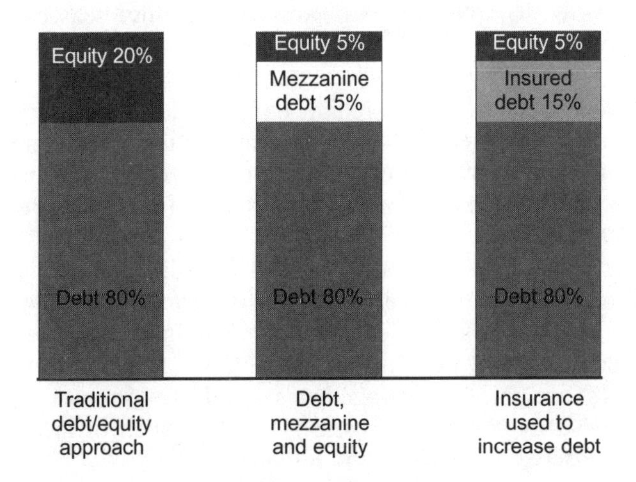

probably need to pay a one-off insurance premium ("mortgage indemnity insurance") to cover the portion of the loan over 75%. Thus the mortgage lender knows that, if he has to sell the house to recover his loan and the proceeds fall short of what he advanced, he is reimbursed by the insurer for losses on the portion of the loan above the original 75% of value.

In the same way, specialist insurers are (or were in the past) prepared to provide indemnities against loss on loans secured on commercial buildings, as protection for the lenders. If banks would normally be prepared to advance only 80% or so of the cost of the project, the insurer might be prepared to provide a guarantee up to 95%. In this case the banks might be prepared to lend up to 95%, as they know that, for the top slice of the loan, they are protected by the strength of the insurance company rather than merely by the security of the property. Again, the insurance is provided in return for a one-off premium payment.

Developers who want to minimise the amount of their own money they put into the project as equity may adopt the insurance route. Or, to be more accurate, they might have done in the 1980s. In the event, some insurance companies lost very heavily on their commercial mortgage indemnity business when the property market crashed, and this kind of insurance became virtually unobtainable, certainly for development loans. It remains to be seen whether it would resuface in happier times (see Chapter 41 for more detail).

Protection against interest rate movements. Since the rate of interest on development loans will normally be linked to LIBOR, the developer (and therefore, in the worst case, the lenders) can be badly hit by a sharp rise in interest rates. The total costs, which include finance charges, rise sharply if LIBOR rises from 8% to 15% during the course of the project. The developer may therefore be required by his lenders to hedge (insure against) the interest rate risk in one way or another — we have already looked at caps and other interest rate hedging techniques. However, this is more difficult on a development loan than on a loan to finance an income-producing investment property, as the loan will be drawn down in stages to meet progress payments on the construction and it is difficult to know in advance what amounts will be outstanding at which point.

Transmuting development loans into investment loans. Many developers ultimately want to build up an asset base for their company, and one way of doing this is to retain rather than sell developments on completion. As we have seen, lenders may be prepared to transmute their development loans into investment loans when the building is completed and let. Again, we are talking of a "normal" period in the property market. Since there were very few development loans anyway in the post-crash period, the question did not often arise in the mid-1990s. But investment loans to finance purchases of revenue-producing properties both were and are available.

The problem both with development loans that transmute into investment loans and with straight investment loans is the old one that rates of interest on borrowings are likely to be considerably higher than the yield from rents on the property, except in unusual circumstances such as those of the 1990-93 market collapse. In terms of capital cover the bank might be prepared to lend two-thirds to three-quarters of value — slightly more generous than the terms for a long-term debenture. But an office building worth £10m and valued on a 6.5% yield is producing rents of only £650,000. A loan of £7.5m, or 75% of value, would carry interest charges of £750,000 even at a 10% interest rate and, as experience in 1989 and 1990 showed, rates can go much higher than this.

Normally, a bank would want to see the interest at least covered by the rental income, and this could impose a lower top limit on the size of the loan. This requirement might be relaxed for a reversionary property where there was a virtually guaranteed increase in income to come within a year or so, but a lender would not normally want to see an income shortfall for a long period. It might also be relaxed at a period of exceptionally high interest rates, when rates were expected to move down before long. Thus, for example, some lenders might allow the company to "roll up" interest above the 10% level (add it to the outstanding amount of the loan), provided this did not take the total loan above the 75% of value limit.

The income cover requirement might be waived if the parent company of the company owning the property was prepared to provide a guarantee for the interest — and was substantial enough for the guarantee to carry weight.

If the loan package for an investment property has been worked out so that the interest is covered by the income, a sharp rise in interest rates could be even more serious than for a development. At least with the development the whole of the loan is not outstanding for the whole of the development period (which, in any case, is likely to be relatively short) whereas an investment loan would probably run for five years or so. So the lenders would almost certainly want the borrower to cap the interest rate on the loan.

In terms of cost, the interest rate margin charged on an investment loan would be lower than on a speculative development loan (the risk for the banks reduces once income is flowing from the property) and could be close to that on a bank facility for a prelet development, or possibly lower. But in the changed conditions of the 1990s, lenders became very much less ready to consider investment loans where the income initially fell short of interest charges. In fact, many lenders (and particularly some overseas ones) based their lending mainly on the cash flow from the property and paid considerably less attention than in the past to loan-to-value ratios.

40
Mortgages for business

The principle of the mortgage has already cropped up in the context of the mortgage debenture: a form of securitised loan secured by way of mortgage on a portfolio of properties. But the mortgage loan itself, in its non-securitised form, is one of the more common forms of property finance.

From the point of view of the lender, the legal mortgage offers excellent security. The mortgagee (the lender) enjoys many of the powers of an owner, but without corresponding liabilities. He will usually have powers to sell the property and recover his loan from the proceeds in a number of circumstances, though in practice he may prefer to appoint a receiver to do the job for him rather than become "mortgagee in possession" and incur the liabilities of ownership.

The circumstances in which the loan might become repayable will depend on the way the mortgage is drawn up, but as well as the obvious cases where the borrower (mortgagor) does not pay the interest or meet the capital repayments on the due date, a number of other situations could trigger repayment. For example, if the borrower fails to observe the covenants (does not insure the property as required, perhaps) or if he is declared insolvent or bankrupt, even where the specific terms of the mortgage have been adhered to.

Many different types of loan may be secured by way of mortgage — site purchase loans and development loans, for example — but we are more concerned with the use of the mortgage as a long-term financing instrument.

And here a distinction has to be made. Mortgage loans are available for smaller commercial property transactions, such as the finance of a corner shop purchase by the occupier. Sometimes the building comprises both commercial and residential space. Transactions of this kind have much in common with the traditional homebuyer's mortgage, and the finance often comes from the same sources: building societies, banks, some specialist lenders and the finance arms of insurance companies (the latter may be using their own funds or may be acting as a conduit for bank money).

With this type of mortgage, the individual loans will not be enormous, the interest will usually be at a variable rate and the loan will probably be repayable in instalments or via a pension plan or an endowment assurance policy.

Mortgages on major investment properties are a rather different market. They were a prime source of finance for the property company in the immediate post-war era, when the insurance companies' life funds looked on property as security for fixed-interest lending rather than as an equity investment that they might own outright. It was common for an insurance company to make a line of mortgage finance, up to an agreed limit, available to the property developer. As buildings were completed and let they would be revalued and the insurance company would be prepared to lend, typically, two-thirds of value at a fixed rate of interest for 25 or 30 years. Before inflation began to bite and forced interest rates up, income from the property would normally comfortably cover the interest on a mortgage of this size. The developer could repay his construction finance with the long-term funds from the insurance company. But, as interest rates rose and insurance companies began to require a stake in the equity of the property, this form of finance became less common.

But some of the life funds do provide long-term fixed-interest mortgages of this type today, and will tend to regard them as just another fixed-interest investment alongside government stocks and company debentures. Where the mortgage is for a property investment company, the security may be particularly attractive. As well as the security provided by the bricks and mortar, there is the covenant of the third-party tenant and, above that, of the borrowing company.

For the insurance company with certain fixed liabilities to meet in future — payments under annuities, for example — a fixed-interest investment with a known redemption date may be attractive. Because the mortgage is not a quoted security the return will be higher than on a quoted debenture — perhaps an advantage of 100 basis points or more — which helps the life fund to increase its average returns.

The disadvantage for the lender is the lack of liquidity. While the mortgage is probably transferable, this does not help if the lender might want to divide the investment among a number of funds in the future. One answer with an individually large mortgage is to insist on "securitising" it. The borrower creates a mortgage debenture on the single property rather than a plain mortgage, and the whole of this

debenture is then subscribed by the one insurance company, which can allocate it between different funds or move the ownership round at a later date if required.

The loan-to-value ratio which the lender is prepared to accept may also be lower on a very large property than a smaller one: perhaps 60% rather than 70%. Loan-to-value ratios will also normally be lower on industrial buildings than on offices or shops. In practice, the size of the loan may be constrained more by income considerations than capital ratios — as with other forms of loan, the lender would not want to see an income shortfall for any length of time.

Normally, the whole of the mortgage loan will be repaid at the end of the term from the sale or refinancing of the property. One major insurance company lender quotes terms of two to 30 years and sizes of loan from £250,000 to the tens of millions. While fixed-rate mortgages are most common, variable-rate may be available. There is also the possibility of negotiating a "droplock". The loan starts off as variable rate, but the borrower has the option of locking into a fixed rate of interest at various points in the future.

While interest on a mortgage can be fixed- or floating-rate, it cannot normally be geared to profits. The reason, as with other forms of debt, is that the interest would not be allowable against tax if the rate depended on the profitability of the business. But methods of indirectly gearing mortgage returns to profits have been devised (see "convertible mortgages" in glossary).

As with a mortgage debenture, the 1986 Insolvency Act poses some problems. The mortgagee, as the holder of a fixed charge, could find he is not able to realise the security immediately to recover his loan if an administrator should be appointed to the borrower. The way round this is again to take a floating charge as well as the fixed one.

There is no limit in theory to the number of mortgages that can be secured on a single property. The mortgages described here will normally be first mortgages, but second and third mortgages might also be created. The lender with the first mortgage naturally has the first claim on the security in the event of default and rates of interest on subsequent mortgages would normally be higher to reflect the higher risk.

One development in the UK commercial mortgage market has been plans to pool and securitise portfolios of commercial mortgages, as has already happened with home loans (see

Chapters 41 and 43). The securitisation process would involve some form of credit enhancement, such as insurance cover for a proportion of the capital value. It is the small business mortgages — corner shops, vets' surgeries, small business units — that are most likely to be securitised in this way. But there has been at least one example of securitising a property company's secured borrowings (see Chapter 41).

41
Credit risk and credit enhancement

All financing arrangements involve some element of risk, but the risk is much higher in some cases than others. A banker who lent 100% of the cost of a development on non-recourse terms would be taking a very big risk indeed — which is why, in theory at least, bankers do not do it.

For example, suppose a development costing £10m and financed entirely on borrowed money turns out to be worth only £8m on completion. It would be possible to repay just £8m of the £10m loans from the proceeds of selling the property. The bank would have lost £2m.

If, on the other hand, the bank lends only 70% of cost — £7m on the project costing £10m — the banker should get his money back even if the project is worth only £8m on completion. It is the people who put up the other £3m — the equity contribution — who lose their money first.

Then look at the same development, again financed with bank loans, but where it is being undertaken by the subsidiary of a very large and financially sound property company. Suppose this company provides the bank with a guarantee for the £10m loan raised by its subsidiary. Even if the development is worth only £8m on completion, the bank should get back the whole of its £10m by requiring the parent company to stump up under its guarantee. The risk for the lender is very different from that in the first case we looked at.

These examples may be simple, but they provide a good introduction to the techniques of credit enhancement: the process whereby the risks to a property lender can be substantially reduced. In essence there are two main forms of credit enhancement:

- A large protective cushion is built into the financing structure, so that somebody else absorbs the first part of any possible losses before the main lender's money is put at risk.
- Some form of guarantee for the loan is provided by a company or body that is financially much stronger than that which is actually borrowing the money. It is much the same as when a well-heeled father provides the bank with a guarantee for the overdraft of his impecunious student son or daughter.

The first type of protection for the lender is — or should be — built into any loan structure. The bank should lend only if the developer has provided a cushion of his own money which bears the initial brunt of any losses.

The second type of arrangement can be used by a property developer or investor either to reduce the price he pays for his finance or to allow him to borrow a higher proportion of the cost of a project than would otherwise be possible — or both. The bank, which sets its interest rate partly by reference to the perceived risk, will reduce its interest charge if the risk is reduced by some form of credible guarantee against loss.

This is why you will find that the very large property investment companies in Britain generally raise corporate loans rather than project loans. The parent company itself borrows the money on the strength of its own standing in the market, though it may or may not offer specific properties as security as well. If a subsidiary borrows money, the parent company may guarantee it. In either case, the cost will normally be lower than if its subsidiaries borrowed direct simply on the strength of their individual development projects.

Property trading companies, on the other hand — even the very large ones — do not generally have the financial strength to follow this course and therefore tended in the 1980s to raise limited recourse loans on the strength of their individual development projects. The money is likely to be more expensive (and given the banks' property loan losses of 1991-93, it certainly should be).

Most property development and trading companies do not have a well-heeled parent that can guarantee their loan or borrow relatively cheap money for them. But it may still be possible to build some element of credit enhancement into their borrowing. Effectively, the company pays a fee to some other body to "guarantee" its borrowings, or at least part of them. The most common example in the 1980s was the "commercial mortgage indemnity" or CMI policy, negotiated with an insurance company.

A CMI policy is not technically a guarantee, but an indemnity against losses. It works in much the same way as the "top slice" insurance on domestic mortgages, which brought very large losses for the insurance companies when residential property values collapsed in the 1990s.

Suppose a developer undertakes a project costing £10m. The banks might normally be prepared to lend 70% of the cost, or £7m, as senior debt, which means that the developer would have to provide £3m of his own money. But the developer does not have £3m. So he arranges to borrow an extra £2m, taking his loan up to £9m, on condition that he provides an insurance company guarantee (technically, an indemnity against loss) for the £2m top slice of the loan. This means he needs to put only £1m of his own money into the project.

He pays a one-off up-front fee for the insurance cover. In return, the insurance company will compensate the lending banks for any losses on this £2m top slice of loans. If the development fetches only £7m when it is completed and sold, the developer loses his own £1m contribution and the insurance company has to stump up £2m to the banks to cover the loss on their loans, so they are protected. If the development fetches only £6m, the banks will themselves lose £1m, as the first £7m of their loans is not covered by the "top slice" arrangement with the insurance company.

This type of policy was popular in the late 1980s, though would have been difficult or impossible for a developer to obtain in the property, market recession of the early 1990s. One insurance company, Eagle Star, made provisions against losses of several hundred million pounds on business of this type that it had undertaken in the late 1980s, and the insurance market is now a great deal more cautious. But it may still be possible to obtain CMI protection for borrowings secured on revenue-producing properties.

Another version of the insurance theme is known as "ground-up" cover. In this case the whole of the loan is covered by an indemnity policy, not merely the top slice. The premium for covering the senior debt will clearly be lower per pound insured than that for the top slice, as the risk is much lower.

Ground-up cover can be particularly useful for syndicated loans — large borrowings where a number of banks are each invited to provide part of the money. Many of the banks approached may have little knowledge of the UK property industry and therefore find it difficult to evaluate a property development financing proposal. But if they view the loan as "guaranteed" by an internationally known insurance company, it is really the credit-worthiness of the insurer that they are required to assess — a much easier proposition.

Thus, in the 1980s, an "insured" loan could often be launched more easily and at a lower rate of interest than an uninsured one.

The danger of any form of loan insurance is that the banks which set up the loan will not bother to evaluate the proposition as carefully as they otherwise would — or will accept a riskier proposition — simply because they know that an insurance company will cover at least part of any possible losses. There is little doubt that the existence of mortgage indemnity insurance led to a great deal of "sloppy" property lending in the late 1980s, both in the commercial and residential field. And insurance companies, and sometimes the lenders, found themselves paying the price in the 1990s.

Insuring a loan or part of it is one aspect of "credit enhancement". But credit-enhancement techniques can also be used in more sophisticated financing structures where there is a need to minimise the risk to the lenders.

Take a mortgage securitisation operation as an example. This is the process whereby a bond is launched in the financial markets, backed by many (perhaps a thousand or more) individual residential mortgages. The lender who originally provided the mortgage loans to homebuyers gets his money back from the sale of the bond to investors. The interest and capital payments from individual borrowers provide the money to pay interest on the bond and eventually to repay the capital amount of the bond. Similar structures have been used — though less frequently — to securitise loans on commercial properties.

The investor who buys a bond backed by individual home loans wants some protection against loss — if a large number of homebuyers default on their loans, it might not be possible to pay the full interest on the bond or to redeem the full amount at the end of the day. This protection normally comes in several stages.

First, the individual mortgages were probably for only 75% of the value of the home or, in the case of a higher loan-to-value ratio, the part of the loan above 75% would have been covered by an insurance company (a mortgage indemnity policy) in return for a one-off up-front premium. Thus, if an individual borrower defaults, the lender does not begin to lose his money unless the house is sold for less than 75% of its value.

In addition, there will be further credit enhancement built into the arrangements for the "pool" of individual mortgages which provides backing for the bond. It takes two main forms.

Under one arrangement, two separate bonds are issued with the backing of the mortgage pool. The main (senior) bond issue might be £100m of floating rate notes — bonds that pay a floating rate of interest and are known as FRNs — which are sold in the financial markets. But in addition, the mortgage securitisation company makes another (smaller) issue of say, £10m of floating-rate notes of a different (junior) class. Thus, the total mortgage pool to back the bond issues needs to be £110m. The smaller, £10m bond issue, pays an appreciably higher interest rate than the main issue in order to recognise the higher risk, because these investors are the first to bear any losses on the mortgage pool. Thus, in our case of a £110m mortgage pool, investors in the main bond issue would not face losses until total losses in the pool exceeded £10m.

Much the same effect can be achieved with the use of insurance cover. Reverting to our example, in this case the mortgage securitisation company makes a single £110m bond issue. But, in return for a premium, an insurance company agrees to cover the first £10m of any losses in the pool of mortgages backing the issue (in practice the figure would probably be a little higher).

There are thus two levels of protection for the investor in the bond. In the first place, the pool would not suffer any losses unless the houses of defaulting borrowers fetched less then 75% of original valuation. In the second place there is the further £10m cushion built into the mortgage pool itself, so that losses to the pool would have to exceed this amount before investors in the principal (or only) bond themselves suffered loss. With the benefit of these forms of credit enhancement, bonds backed by domestic mortgages can achieve the highest credit rating (though they are not quite so popular since the house market in Britain went into a nose-dive at the end of the 1980s; and with the two-tier bond structure it may be necessary to provide some insurance on the junior bond).

It goes without saying that the insurance company providing the cover for the loans must itself enjoy a very good credit rating. In the United States there are a number of insurers specialising in this kind of financial guarantee business, and their operations extend to the UK. As with other forms of insurance, they work on the principle that only a proportion of the risks which they underwrite will result in claims.

A similar credit enhancement structure was used by a British property company called BHH in 1990. In essence, BHH sold £90m of five-year floating rate notes in the investment markets. The notes were backed by a portfolio of commercial mortgages secured on £83.6m-worth of BHH's properties. If this looks rather a high loan-to-value ratio, BHH actually received only £62.5m of the proceeds, the remaining £27.5m remaining on deposit but available against approved future property purchases. The loan-to-value ratio was thus 75%.

Therefore, there was effectively a cushion of equity in the individual mortgages to provide the first line of defence against loss. In addition, an insurance company called Financial Security Assurance or FSA provided an unconditional guarantee for the whole FRN issue. And in turn FSA laid off part of its risk with another credit risk insurer which agreed to bear the first 20% of any possible losses. With the benefit of this level of credit enhancement, BHH's FRN issue initially earned the highest rating from the credit agencies who assess risk on behalf of investors.

A more recent financing with some similar features involved 135 Bishopsgate, part of the Broadgate development in the City of London which was undertaken by a company owned equally by two property developers, Rosehaugh and Stanhope. This particular building in the development was occupied by National Westminster Bank, which counts as a tenant unlikely to default on the rent.

It was decided to finance the building by making use of the commercial paper market in the United States. This is a market in which companies can issue short-term "IOUs" that usually have a life of only a month or so to raise cash. The idea is that, as the original IOUs are repaid, fresh ones can be issued so that the market can, in practice, provide a longer-term source of funds. But as a precaution, in case it ever proved impossible to "roll over" the short-term debts in this way, the issuer of the commercial paper also lines up a "standby facility" or "liquidity facility". Under this arrangement a group of banks agrees to lend the funds if the company is unable to raise cash by selling further IOUs to investors.

Via a complex swap arrangement which need not bother us, the rent paid by National Westminster provided the interest on the IOUs which were issued in the US commercial paper market. Investors in this market knew that the IOUs could be repaid because of the liquidity facility which would provide the funds if necessary. But as protection for the investors, and also for the banks providing the liquidity facility, the whole of the $178m borrowing was unconditionally guaranteed by FGIC, a US financial guarantee insurance company. With a loan-to value ratio between 60% and two-thirds, the borrowing arrangement thus had several layers of credit enhancement built in, resulting — it was claimed — in lower borrowing costs for the property owner.

But nothing stays still for very long in finance. This arrangement has, as it happens, been refinanced subsequently by a new owner of the property.

42
Leasebacks old and new

The most popular off balance sheet financing structures of the late 1980s boom involved the use of a company that was not technically a subsidiary of its real parent. This off balance sheet company borrowed for property development or purchase and its borrowings did not appear in the group balance sheet of its parent.

But there is an older form of off balance sheet financing structure that does not involve borrowings at all: the leaseback or sale and leaseback, which we have already briefly mentioned. It would be unusual to find leasebacks used to finance developments nowadays, but leaseback transactions entered into in earlier times may still be found in some property companies. And the leaseback is still extensively used as a method of raising finance by businesses which own large amounts of property as an incidental aspect of their business: stores groups in particular.

Suppose an efficient and aggressive stores group — call it Pilem High Holdings — acquires a less successful competitor: Stodgy Stores. Stodgy has been earning a poor return on capital and its return would be lower still if the shops from which it operates were included in its accounts at a realistic up-to-date valuation. The only thing that has kept Stodgy going is the fact that it owns the freeholds of most of its shops and therefore does not have to pay rent.

After the takeover, Pilem High reckons it can greatly improve Stodgy's returns. It calculates that the sums of money tied up in owning the shops can be put to better use in the retailing operation itself. But it still needs the stores to operate from.

Pilem High therefore looks for an investor — probably an insurance company or pension fund, but possibly a property company — that would be prepared to buy the Stodgy stores as an investment. But, as part of the deal, the buyer must grant Stodgy a lease of the properties.

After the deal, Stodgy becomes a tenant and pays a market rent for its stores. And it has released cash which can be used in the expansion of the business or to repay expensive borrowings. In the short run at least, the operation should improve its earnings.

Take an example. The Stodgy properties have a current market value of £50m. They are sold to an institution at this price. Stodgy then pays a market rent which is equal to, say, 6.5% on the value of the properties: an annual rent bill of £3.25m. But with the proceeds of the sale it can repay £50m of bank borrowings which were costing it, say, 12%. It thus eliminates £6m a year of interest, giving it an immediate saving of £2.75m after knocking off its rent costs.

In practice the calculations will be a little more complex than this, partly because of the costs involved in the transaction but more particularly because of the tax aspect. Stodgy will probably be liable to tax on the gain it makes in selling its properties at £50m, so its net receipts will be less than £50m and the actual savings on interest will be somewhat lower than in the example. But in many cases the transaction will still be attractive. It is a particularly common move after leveraged buyouts in the retail sector, where the new owners need to realise cash as soon as possible to reduce the level of debt incurred in the buyout.

The disadvantage of the leaseback is that the original owner loses any further interest in the growth in value of the property. His rent will be revised upwards at the normal intervals and what looked like very cheap funding in the short term may turn out to have been considerably more expensive when he faces the higher rent bills. For this reason an off balance sheet variant of the sale and leaseback theme may sometimes be preferred. A 50-50 joint-owned company is set up which borrows the bulk of the money to buy the properties, granting the vendor a lease in the normal way. Via its half ownership of the joint company, the original owner retains an interest in the performance of the properties in the future.

An alternative, particularly if the vendor does not want to be faced with a very large rent bill, is for him to sell the properties to an institution at below market value, and pay below market rents in return. The properties might be sold for three-quarters of full value, with the tenant paying a rent three-quarters of the market level. The capital proceeds from the operation are lower, but so are the subsequent outgoings, and effectively the vendor retains a minority equity interest in the property.

Even where this factor does not intrude, terms of a sale and leaseback transaction will not always be taken as a reliable indicator of the property market in general. The vendor may be prepared to concede a rent slightly higher than he would do in an arm's length transaction in order to ensure the success of the deal, reckoning that any disparity would be sorted out at the first review. Sometimes he also agrees to guarantee minimum uplifts in the rent at the review point, regardless of market conditions, which can be an important additional benefit for the purchaser.

The sale and leaseback used in development financing in the past works slightly differently. There were numerous variants, but an example conveys the principle.

A developer identified a site for an office block. If possible, at the outset he persuaded an institution to buy the site and provide the finance for construction and other costs and the developer undertook the project. In return, the institution granted the developer a very long lease of the completed building.

Suppose the all-in cost (including site) was put at £10m, and that Myopic Mutual Insurance agreed to provide this £10m to finance the office development proposed by Payola Properties. Myopic Mutual stipulated an initial return of, say, 7% on its investment or £700,000 a year. Above this level the rent might be split 50-50 with the developer.

If the developer let the building at £1m a year, or 10% on cost, he would pay a ground rent of £850,000 to the institution under the terms of the lease and keep the remaining £150,000 "top slice" as his own profit rent. If he managed to achieve a rent of £1.2m, the institution would get £950,000 and the developer £250,000.

The proportionate division of the cake at the first letting stage would probably be maintained for the life of the lease — though more and less favourable deals for the developer were sometimes negotiated. If the institution got 70% and the developer 30% of the rack rent at the outset, these proportions would thus usually remain constant as the rack rents rose. So the ground rent the developer was obliged to pay to the institution would rise at each rent review on the building. If, in our example, Payola Properties suddenly lost the tenant of 40% of the building and had problems in reletting, its rent to the institution would exceed the income from the property and it would have to make good the difference.

Developers used leaseback finance as a way round some of the tax problems of a joint-venture company and because it provided finance at rates below those for borrowed money — though at the cost of giving away a great deal of the future growth. The very real future liability to pay rent to the institution did not appear in the balance sheet as a borrowing would have done, so the liability was not always obvious.

The developer's "top slice" income was not awarded as high a capital value as rents from a wholly-owned building. Subsequently, there has been a trend to unscramble many of these earlier leaseback deals: the institution might buy out the developer's profit rent, or the developer buy the institution's superior interest. As an undivided investment, the building would be worth more than the sum of the institution's and the developer's interests, so there was profit to be made by bringing the two interests back together and realising this "marriage value".

A new development of the late 1980s and early 1990s was the "tax-driven" leaseback, particularly popular for hotel properties or properties in enterprise zones where large capital allowances were available. The property owner "sold" the property to a financial institution and took a lease back at a "rent" geared to money-market interest rates rather than property returns. The financial institution could use the property's capital allowances to reduce its own tax bill, and some of this saving was passed on in the form of lower "rent" to the original owner. This original owner almost certainly had an option to buy the property back later, effectively at the same price as it had sold it for. The transaction was frequently structured to take the property off the original owner's balance sheet.

Such financing transactions have not disappeared, but by the mid-1990s their accounting presentation looked rather different. Accounts now had to reflect the substance of a transaction. And the substance of the tax-driven leaseback was that the property had not been sold at all, since the original owner had to option to take it back at the same price. All that had really happened was that the owner had raised a secured loan against his property, with some interest savings through the tax angle. And that was how the transaction had to be presented, with both the property and the associated "loan" shown on the original owner's group balance sheet.

43
Unitisation and securitisation

The search has been on for several years to find a way of "unitising" properties. What this means is that the ownership of a property or a portfolio of properties would be spread among a number of investors. Instead of owning the property or properties outright, the investor would own a piece of paper that gave him the equivalent of ownership rights to a portion of the property or properties.

Unitisation is one aspect of "securitisation": another popular buzz-word of the 1980s and 1990s. Securitisation is a form of financial engineering that converts underlying investments or assets into "securities": pieces of paper that can be sold and traded in a market. A prime example in the UK is residential mortgages. Some mortgage lenders to homebuyers put a thousand or so of these mortgages into a "pool", then sell bonds or other forms of tradable loan in the securities markets, backed by the pool of mortgages.

The interest paid by homebuyers on the mortgages, and their repayments of principal, provide the income for the bond and the capital with which ultimately to redeem it. Meanwhile, the mortgage lender has recouped his original outlay from the proceeds of sale of the bonds and released funds with which to make further loans to homebuyers.

There are several reasons why the financial and property communities are keen to unitise or securitise properties. The market in physical properties is relatively cumbersome and illiquid. Transactions take a considerable time and there is a limited number of buyers with millions of pounds to invest in a single property. But if pieces of paper conferring ownership rights to property could be traded in a market instead of the properties themselves, transactions could take place very much more quickly and a far wider range of investors could be tapped. This would be particularly useful in the case of very large individual properties.

Then there is the problem of spreading risk and obtaining the benefit of professional management. Even among institutional investors there are many smaller concerns — medium-sized pension funds in particular — that might like to invest in commercial property but would be reluctant to spend several million pounds on a single building. It would involve putting too many eggs in one basket, and they might not have the property expertise to manage the investment effectively.

The efforts to unitise or securitise properties have thus been aimed in two main directions:

- There have been attempts to divide the ownership of very large single buildings, so that a wide range of investors could each have a stake in a single building.

- Other initiatives have aimed at providing professional management and a spread of risk by offering investors a stake in a portfolio of different properties.

Take the very large individual building first. These properties can be particularly difficult to sell. Even the big insurance companies and pension funds think twice before committing as much as £50m to a single building. One result is that these very large buildings are often valued at a discount to otherwise similar but smaller investments. If a £10m office block is valued on a yield of 6.5% (at 15.4 years' purchase), a £50m block of the same quality in a similar location might be valued on a yield of 7% (at 14.3 years' purchase).

The discount varies with market conditions. When there are a lot of big buyers around, the discount may be greatly reduced or may almost disappear for a time. But it is generally present. From this has grown the idea that, if ownership could be unitised among a range of investors, the valuation discount could effectively be removed. In other words, the units would be more liquid than the property itself and the aggregate value of all the units might be greater than the value of the property as a single entity.

A number of different routes have been followed in the attempt to unitise large individual buildings. The problem throughout has been that it is very difficult under English law to divide the ownership of an individual property.

One route was the Single Property Ownership Trust or SPOT. The trust would own the property and, via the trust, this ownership would be spread among a range of investors on the unit principle. But for the scheme to work it required "tax transparency". In other words, income from the properties (and profits on disposal) must be able to come through to the unitholders as if they owned the property direct, without prior deduction of tax. The tax authorities would not concede this principle and the idea was put on the back burner.

Another approach involved Property Income Certificates or PINCs. Via a complex structure of companies and leases, the benefits of ownership of a single property (ie, rental income and any capital gains) could be apportioned among a range of investors, who would receive their share of the income (less management and other charges) as if they owned a corresponding stake in the property direct.

The idea was that the PINCs would be traded on the Stock Exchange and market forces would determine their value. Thus PINCs investors would not directly own a proportion of the building, but they would own securities that conferred comparable benefits. But again the idea was put on the back burner when the UK property market as a whole became more nervous in 1989.

A third approach involved a company structure, but the company — Billingsgate City Securities was the only example up to the mid-1990s — owned a single property: a large City of London office block. In this case the company — a single asset property company or SAPCO — financed its ownership of the property with three different classes of security: a deep discount bond, preferred shares and ordinary shares. Only the preferred shares — whose income rose with increases in rent on the building, but which did not have any benefit of the gearing — were available to the public and they did not prove popular. The preferred shares were ultimately bought back and the building was eventually sold to its tenant.

Thus the idea of providing greater liquidity in the property investment market, by converting ownership of individual properties into ownership of more liquid securities, has not so far had much success.

On the face of it, the property company structure might seem the most natural one for securitising either an individual building or a portfolio of properties. Indeed, the stock market's listed property companies are, in a sense, an example of securitised property. Rather than owning properties direct, the investors — shareholders — own shares in a company that in turn owns the properties.

But there are snags. Though the value of the shares will to some extent reflect the value of the underlying properties, it will also be affected by a lot of extraneous factors. The value of property company shares will probably fall when the stock market as a whole is in decline, even if the value of the underlying properties has not been affected.

The tax position may be a still greater impediment. The mainstream corporation tax that is deducted from the company's income cannot be recovered by investors, and the company will be liable to capital gains tax on profits on sale of its properties. So in tax terms, owning property company shares is not a proxy for a direct stake in a property.

As an alternative to the property company structure for owning a portfolio of different properties, unitisation has had some success. The principle of "pooled" or unitised investment in this case is relatively simple. Suppose a thousand investors put up £10,000 each, and the resultant £10m is invested by professional managers in a range of properties which provides a spread of risk. The value of each of the 1,000 units rises or falls to reflect movements in the values of the underlying properties and each unitholder receives the appropriate share of the income from the properties, after management expenses.

It was not possible in the past in Britain to operate an "authorised" unit trust that owned properties direct, let alone a single property, though authorised unit trusts owning a spread of properties are now permissible. The climate of the 1990-93 property market collapse was not, however, favourable to their creation.

By 1996 the London Stock Exchange was planning to allow the listing of units in these authorised property unit trusts (or APUTs). This would mean that the units could be traded in the market and investors would not be limited to buying them from (or selling them to) the trust's managers. It was hoped that this would lead to greater liquidity and allow this type of vehicle (somewhat belatedly) to take off. But the APUTs would not – without further concessions – be fully tax-transparent. In other words, income from the properties owned by the APUT would not come through to the unitholder without deduction of tax. Without this feature, the cynics argued, the attractions of the APUT vehicle would remain limited.

Although property unit trusts could not, until they were authorised, market themselves to the general public, there had been a long-standing dispensation for unauthorised property unit trusts whose units could be bought only by tax-exempt funds. This meant mainly pension funds and charities. These unauthorised property unit trusts offer the benefit of professional management and a spread of properties to smaller tax-exempt investing institutions. The value of the units depends on the value of the underlying properties that the trust owns.

For the private individual, "property bonds" have offered some of the same characteristics, though the legal structure is different. Theoretically, these are a form of unit-linked life assurance. But the value of the life assurance contract depends on the value of the units, which in turn is determined by the value of a portfolio of properties the life fund owns.

Despite the problems encountered in unitising individual properties, the idea of securitisation as a funding method is far from dead. Deals were undertaken or planned which embodied the principle, in the days before stricter accounting rules probably prohibited an off balance sheet treatment. Suppose Company A wanted to raise money on the strength of its properties or a single property but keep the debt off its balance sheet. A special purpose vehicle (not a subsidiary) could have been set up which borrowed to buy the properties from Company A. In turn, it leased the properties back to Company A for its own use.

Company A paid rent to the special purpose vehicle, and contracted in advance to pay certain increases at the review periods. Insurance was arranged to underpin the value of the properties and, with this protection and the guaranteed rent increases, a high proportion of the value of the properties could have been borrowed by the special purpose vehicle, even if rents did not initially cover interest charges. Debt could have been in the form of bank loans or the special purpose vehicle could have sold debt securities to investors, in which case it had effectively "securitised" part of the value of its properties. The ingenuity of property financiers knows no bounds, but the new accounting rules which would almost certainly require the borrowings to be shown on the balance sheet in a deal like this may have dampened their ardour.

After all the attempts at securitisation of properties over a period of years, it came as something of a surprise when, in 1994, a "derivative" form of property investment was eventually launched by Barclays Bank. It was not, strictly, a securitisation. But it offered large institutional investors a way of investing in the fortunes of the property market without the need to go through the hassle and expense of buying and managing properties.

The new investment was called "Property Index Certificates" or PICs. Basically, large investors could buy from the Barclays group a form of bond whose returns depended on the performance of the property index run by Investment Property Databank or IPD. This index reflects the performance of the bulk of the investing institutions' property holdings.

The investor in PICs is paid annual "interest" that reflects the average rental yield on the IPD index, less a smallish deduction for Barclays. The repayment value of the bonds — which were issued for two, three, four or five years — depends on the movement in the capital value of the IPD Index over the period. Suppose an investor bought £1m of three-year PICs and over the three years the value of the IPD index rose 20%. He would be due to receive £1.2m on redemption.

Though no specific properties were securitised, Barclays presented the PICs as a way of hedging the risk in certain of its own property loans. If property values went up, it would have to pay out more on redemption of the PICs but its recoveries from its own property loans would also have risen. If property values fell, it would recover less from its own property loans but likewise it would gain on the PICs because it would redeem them at less than they had originally been sold for.

Subsequently, the Barclays group also introduced Property Index Forwards — an over-the-counter derivative product that allowed investors to bet on, or to protect themselves against, movements in property values in the future. And by 1996 a group of heavyweight City of London institutions was pursuing even more ambitious plans for a traded market in property derivatives under the title of the Real Estate Investment Market or REIM. With the earlier fiasco of London FOX now largely forgotten, the hunt for viable property derivative products was once again in full cry.

44
How property bucks the trend

In theory, returns on different kinds of investment bear some relationship to each other. If yields on government stocks rise, all else being equal investors will probably expect a higher return on ordinary shares as well, to maintain the relationship between the two different investments.

The theory should apply to commercial property as well as to stock market investments. Suppose you can get a completely safe 10% redemption yield on long-dated government stocks. And assume that at this time the average yield on property is 6%. The property investor is obviously making some assumptions about growth. He might be assuming, say, that income from his property (and its capital value) will grow at 8% a year. Taking into account the effect of the five-year rent review pattern, this means he is looking for an overall return of about 13.2% from the property. For the purposes of the example, we will call it 13%.

If his assumptions are right, this gives him an overall return 3 percentage points higher than he could get on the government stock, which he may feel is fair compensation for the greater risk and uncertainty of investing in property rather than in gilt-edged securities.

If interest rates in general should rise and the redemption yield on government stocks climbs from 10% to 11%, we might expect the investor would then require a higher yield on his property to maintain the 3 percentage point safety margin between the return on property and the gilt yield. In other words, values of property would need to fall to the point where the income yield to a buyer rises from 6% to 7% to give about a 14% overall return.

So much for the theory. Unfortunately for the theorists, it is not what has usually happened in practice. Look at the charts of the last 16 years or so. Far from moving to reflect changes in returns on other forms of investment, commercial property (certainly prime property) in Britain has usually followed a pattern of its own which appears to owe very little to what is happening elsewhere on the investment scene.

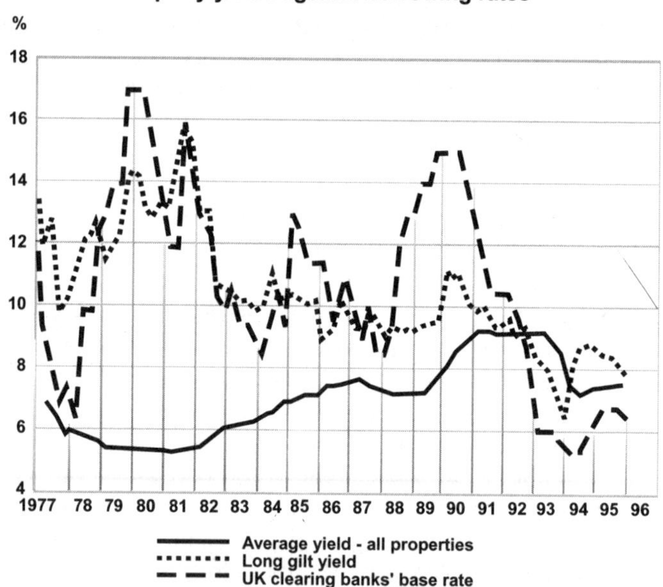

For much of the post-war period, property yields were well below borrowing rates. It required ingenuity from developers and investors to get round this problem, as the income from the property would not cover the interest on the loans raised to buy it. The position changed dramatically in the early 1990s when interest rates at last began to fall back from their turn-of-the-decade peak but property yields remained historically very high. The graph plots property yields against the long gilt yield (an indicator of longer-term borrowing costs) and bank base rates (an indicator of the cost of bank loans). But remember that a commercial borrower such as a property company will pay a margin over gilt yields or bank base rates. Source: *Hillier Parker; Datastream*

It does not matter whether we take short-term interest rates (bank base rates) or yields on long-term government stocks for our comparison. Prime or even average property yields have often tended to move in precisely the opposite direction to these common yardsticks. Secondary property, however, where rental yields are higher and probably provide a greater proportion of the overall return than they would with prime property, may be somewhat more responsive to interest rates.

In 1978 and 1979 average yields on property were moving down while interest rates were rising very sharply. Then in 1982 and 1983 property yields began moving up while interest rates generally were falling. And in 1988 average property

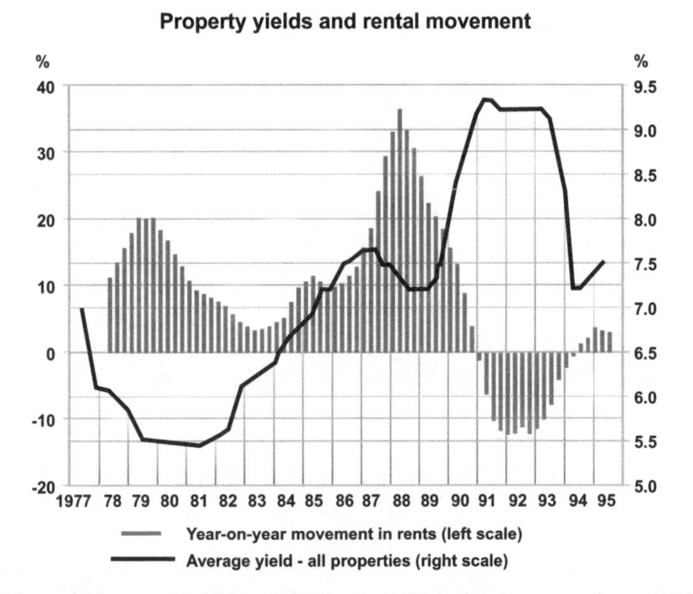

You would expect yields to be lower when the prospect for rental growth is good and to be higher when rental growth is poor or non-existent. Investors will accept lower initial returns when they expect future income to increase rapidly. Indeed, this is the usual pattern, though the graph shows that yields are sometimes slow to adjust to changes in expectations of rental growth. And the period 1984 to 1987 was a curious one, with yields rising despite accelerating rental growth. Did investors have a premonition of the problems to come at the turn of the decade? Source: *Hillier Parker; Datastream*

yields were finally moving down again just as both long- and short-term interest rates were moving up.

By the end of 1989, considerably after the main interest rate hike had taken place, property yields were again moving up to reach a peak in 1991. And it was not until the second half of 1993, considerably after short-term interest rates had begun to fall with sterling's exit from the ERM, that property yields also began to fall again.

However, in the latter part of 1993 and the early part of 1994 we did have one clear example of the influence of interest rates on property yields. During that period returns on bonds and on ordinary shares had fallen very low. In their quest for a better income, investors turned to property and their buying began to move prices up and yields therefore fell. It was, however, a straight comparison between income yields. At that point investors were expecting little if any growth from property.

With this exception, we might deduce from the pattern of recent years that property has its own internal logic which causes it to move in exactly the opposite direction from other forms of investment. But this would be an over-simplification. Logically, one would expect yields on property to edge down as expectations of rental growth increase — investors are prepared to accept a lower initial return in exchange for the prospect of greater growth in the future.

But look at the chart of percentage rental growth over the past 16 years or so. This shows, on a quarterly basis, the rate of increase in rents over the corresponding quarter of the previous year. In 1978 rental growth was rising sharply. It then flattened out and fell sharply from 1980 to 1983. But it was not until 1982 that average yields began to move up significantly to reflect lower growth expectations. They then rose continuously to a peak in 1987, though rental growth had begun to accelerate again very sharply some time before this.

The period from 1986 to the turn of the decade shows an unusual pattern. Despite their modest dip in 1987 and 1988, from the middle of the decade property yields had been at considerably higher levels than might have been expected, given the evidence of very rapid rental growth. With the benefit of hindsight, it perhaps suggested that the market regarded the rate of rental growth as suspect and was looking towards troubles ahead. If this was the case, the market was proved absolutely right by the tumbling rental values of the early 1990s.

From the pattern of the past decade-and-a-half we can deduce that property yields do react to some extent to expectations of rental growth from property. But they usually react with considerable delay and in a pretty haphazard manner. We have to conclude that the market in investment properties is a very imperfect market.

What determines the cycle of rental growth in property? We have to be careful of generalising because City of London offices or prime shops, say, may follow a very different cycle from Midlands industrial buildings. But, as in most markets, it comes back to supply and demand. The level of development activity can obviously affect the supply side of the equation. And the general level of activity in the economy has a strong bearing on the demand.

By 1980 Britain was in recession, the manufacturing sector was particularly hard hit and unemployment was beginning its climb to over the 3m mark. Fewer workers in employment meant less demand for the buildings in which they would work and depressed profits limited companies' ability to pay higher rents. Demand for industrial property was particularly badly hit. When economic activity picked up again, it took some time for surplus space on the market to be absorbed before emerging shortage began to force rents upwards.

Then a new development boom — particularly in City of London offices — emerged in the latter part of the 1980s to capitalise on the higher rents. But by the time much of the new space came on the market in the early 1990s, the country was in recession. A rise in supply and a fall in demand coincided to deliver the worst property slump since the war.

And this is where we must start looking for the cycles in property: at the level of supply and demand. This in turn affects rental growth which, in a very imperfect way and often distorted by temporary enthusiasms or lack of enthusiasm towards property investment, affects yields and capital

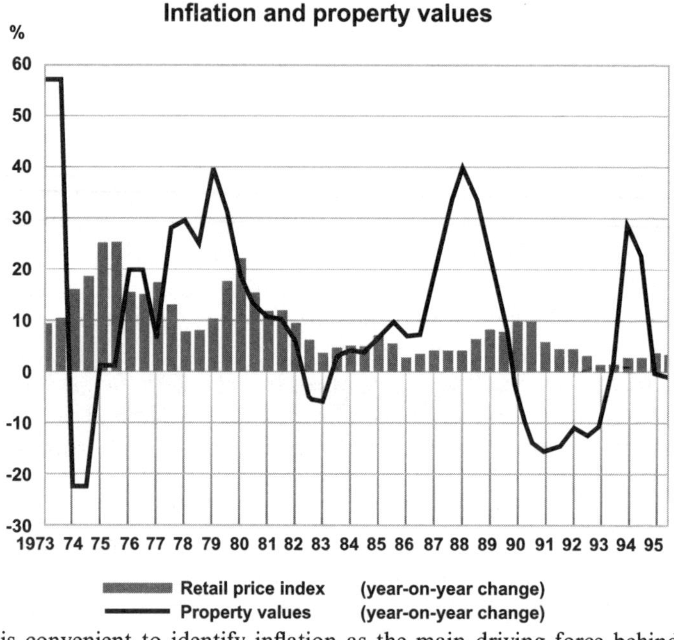

It is convenient to identify inflation as the main driving force behind growth in property values. In the short- or medium-term it is also often wrong. Our chart shows the annual rate of growth in retail prices (shaded bars) and the annual rate of movement in commercial property values. The periods of highest inflation have by no means necessarily coincided with the periods of greatest property growth. Source: *Hillier Parker; Datastream*

values. Don't look for any pattern in property yield behaviour — certainly not in prime yields — more precise than this.

Don't, above all, fall for the old myth that property values rise largely as a result of inflation. In the very long run inflation does have an effect on the value of property, partly because building costs rise in nominal terms and industrial and commercial companies' profits also rise in nominal terms, thus enabling them to meet higher rents. But some of the best gains in rents and property values have taken place at times of relatively low inflation by British standards, as in 1987-88. And the great property crash of 1974-75 took place while inflation was at its highest level this century.

45
Trends in finance and property cycles

An historical perspective is no bad thing when evaluating property financing techniques and investment fashions. To complement our examination of today's finance and investment markets, here is a quick review of the post-war property booms — and busts. This chapter takes us up to 1990. It is followed by a separate chapter on the crash of 1990-93, which rates individual treatment.

The post-war development boom. In the immediate post-war period, yields on property were generally higher than long-term borrowing rates and there was no problem in financing, say, an office block with long-term mortgage funds, which were available from the insurance companies in particular. Construction finance would probably have come from the banks or sometimes from the contractor. The developer's aim, as today, was to use other people's money as far as possible.

Suppose a developer put up a building at an all-in cost of £1m and let it at a rent of £100,000 a year. If the building was valued at 15 years' purchase (on a 6.7% yield), it would be worth £1.5m as an investment. The developer could normally raise a mortgage of two-thirds of value, so he was able to recoup the whole of his development costs and did not need to leave any of his own money in the venture. He had created an equity value of £500,000 for himself (the £1.5m value less the £1m costs).

If the rent was below £100,000, the building would be worth less than £1.5m, the developer would not be able to recoup all of his costs via the long-term borrowing and he would need to tie up some money of his own in the venture. If he achieved more than £100,000 in rent, he would be able to borrow more than £1m and would actually be generating cash. The name of the game, needless to say, was to avoid leaving your own money in the project.

Most of the post-war entrepreneurs held on to their completed developments rather than selling them for a trading profit. This was partly a matter of the tax regime. Tax would be payable on trading profits, but there was no tax on the surplus created by a development until the development was sold.

In due course, many of these entrepreneurs launched their companies on the stock market in the later 1950s and early 1960s. By selling part of their shares, they could realise some cash without needing to sell the buildings themselves and meet a tax bill.

The early post-war developers liked to let their buildings on leases as long as possible, often without review. They were thinking of the profits made from the development operation, and subsequent inflation of property values did not figure large in their deliberations. In due course the institutions which provided the long-term mortgage or debenture funds began to cast an envious eye on these development profits and insisted on a share: sometimes by way of conversion or subscription rights into shares, sometimes via joint company structures for the developments.

By the late 1950s or early 1960s, inflation was beginning to rise and long-term interest rates were rising with it. This reinforced the trend towards profit sharing with the financiers. Once the rate of interest on long-term money exceeded the yield on quality properties (a position that persisted into the early 1990s, though was briefly reversed in the crash of that era), developers had a problem if they wanted to retain their developments on completion. They were forced to give away part of the profit in return for finance at lower rates. Sometimes this continued to be done by incorporating conversion rights into long-term debt. But increasingly, the lease-back became the popular method for insurance companies and pension funds to finance developments.

On the stock market, a long post-war boom in property company shares came to an end in 1962 as profitable development opportunities decreased. In 1964 the developer's life was made still more difficult by strict controls on new office building. Paradoxically, this helped to create shortages which pushed up the rents for existing buildings.

The financial boom of 1967-73. By 1967 property shares were coming to life again as it was realised that a combination of shortage and inflation had brought about significant increases in rental levels, particularly for London offices. The assets of the existing property companies were becoming immensely more valuable, more because of the effects of shortage on their existing holdings than because of any further development surpluses they were able to create.

Property company takeovers became commonplace as the more actively managed companies gobbled up the sleepier ones for the value of their assets. Recognising the reality of inflation, rent review periods had fallen first to 14 years, then to seven years and rather later dropped to today's five years.

There were some quite sharp ups and downs, but by and large property shares remained in favour till late in 1973. Rising inflation was swelling the cash flow into the institutions, which began to buy property on a larger scale as an inflation hedge. They also helped to finance an enormous development boom in the early 1970s, frequently by providing leaseback funds. The other major providers of finance were the banks, whose lending for property shot up.

Many developers borrowed short-term for their developments, or to finance acquisitions, reckoning that rising asset values would more than counterbalance the cost of deficit financing techniques. They relied on borrowing against the increased value of their properties to meet the income shortfall, completely ignoring the importance of cash flow.

Property share crashes

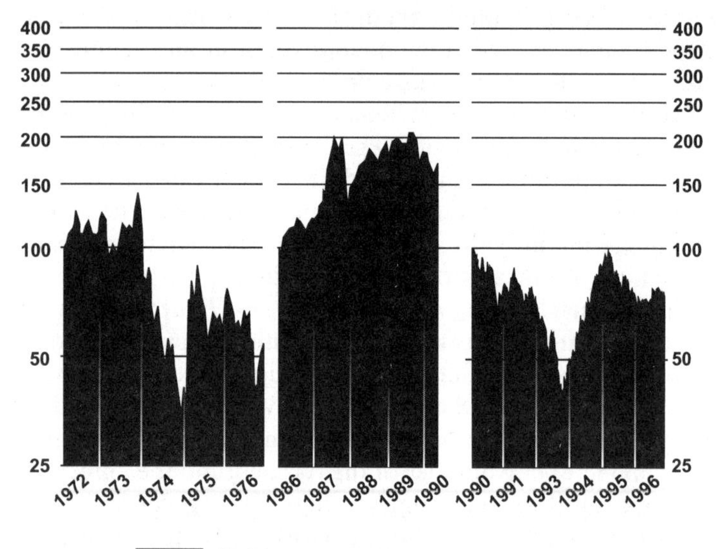

FT-SE property share price index (rebased to 100)

The past 25 years have seen three major crashes in property shares, compared here on a logarithmic scale to emphasise the percentage falls. The crash of the mid-1970s was the most severe, partly because it accompanied a general stock market collapse. With the perspective of time, the crash of autumn 1987 – again part of a wider stockmarket crash, and dramatic at the time – was a mere fleabite. The property share crash of the early 1990s was considerably more serious and largely reflected conditions that were specific to the property sector, rather than a more general stockmarket slump. Note that the starting point for each graph has been rebased to 100. The three graphs do not provide a comparison of the absolute levels of property shares in the three periods. Source: *Datastream*

The end of the boom in asset values and of the finance-fuelled development boom came late in 1973 when, in the aftermath of the oil crisis, interest rates were raised to penal levels. At the same time, secondary banking lenders who were heavily committed to property began to collapse.

The accounting conventions of the time had disguised the sharply negative cash flow of most property companies with large development programmes. Not only was borrowing now prohibitively expensive, but property values were plummeting. It became almost impossible to sell property or borrow further against it (and very difficult to value it). Property shares collapsed, as did the direct market in properties.

The next few years were a period of retrenchment when most of the major property companies were fighting for survival, and a few went under. The pension funds and insurance companies, with swelling coffers, had their arms twisted by the Bank of England to help the banking and property sectors by purchasing property investments from distressed property companies. With the benefit of hindsight, they picked up many bargains.

An investment boom fuelled by institutional buying. By 1977-78 the much slimmed-down property companies were getting back on an even keel. The institutions were still keen buyers of property and values rose sharply in the period 1978-1981. By now, an increasing number of the big institutions were undertaking their own developments or financing developers on what were effectively project-management terms (the developer had a stake in the project, but the institution could buy him out at the end of the day). However, the larger established property companies with spare cash flow could still borrow to develop, hoping that rising rents would eliminate any income deficit by the first rent review.

The late 1980s: a rental boom and a finance-led development boom. By 1982, recession and rising unemployment were hitting tenant demand for property and the rate of rent increase dropped sharply. At the same time yields were increasing to reflect the deteriorating prospects. The property world took a lot longer to recover than the economy itself, and it was not until late 1986 or early 1987 that another great boom in rents and values got under way.

By then the institutions had become less important as long-term finance providers for property (they were buying significant quantities but selling a lot as well, as they took the opportunity to weed out their portfolios). Thus net purchases were lower than in the past. Another development boom was under way, but financed mainly by the banks, which were prepared to lend on the fortunes of individual developments and roll up interest during the development period. The idea was to recoup their money when the property was sold at the end of the day. By definitition, many of the developers were traders rather than investment concerns.

But by early 1990 rental growth had tailed right off, property yields were rising and the market was becoming very sticky. With the institutions unlikely this time to bail out property companies, many banks were reconciling themselves to extending their development loans into investment loans for a further five years or so and committing themselves to the property market for far longer than they had originally envisaged.

46
The crash of early 1990s

And that was how it looked in the summer of 1990 when the original edition of "Property and Money" went to press. Trouble looming, but the depth and duration of the property market recession still far from clear. The commercial property market was, after all, accustomed to cyclical downturns. But something of a different order was on the way.

Over the following three years, many of the assumptions on which the post war property market was based were shaken to their foundations. As the economy as a whole sank into recession, business confidence evaporated. Interest rates, which had begun moving up sharply in 1988, stayed at historically high levels as the government sought to squeeze inflation out of the system and — increasingly desperately and ultimately unsuccessfully — to keep the pound sterling within its exchange rate bounds in the European Monetary System.

The property industry suffered worse than most areas of the economy for three main reasons. First, the asset price inflation of the late 1980s left property values very high and exposed to any downturn in sentiment. Second, the accompanying development boom had ensured that much new space was coming on the market just at the time that prospective tenants were trimming back their own businesses and avoiding any new commitments. Third, the development boom had been financed mainly on borrowed money, leaving developers seriously over-geared and especially vulnerable to a prolonged period of high interest rates. To understand what happened next, we need to look first at the direct consequences for the property world, then at their implications. In the following chapters we examine some of the problems thrown up by the 1990-93 property crash. But first, let us take some the main events in order:

- The combination of increased supply of space and shortage of tenants made it increasingly difficult to let new buildings. The effects were most marked in the London office sector, but were felt in most parts of the country.

- Developers with empty buildings had no income to pay the interest charges on their development loans. These charges were, in any case, now much higher than originally envisaged.

- The development and trading companies needed to find tenants before they could sell their buildings to long-term investors, recoup their development outlay and repay their bank loans. Without tenants the properties were virtually unsaleable and the bank loans could not be repaid. The banks began to close in. Some companies were put into receivership straight away. Others — known in the business as the "living dead" — were kept alive on a care and maintenance basis, though later the banks often lost patience and closed them down.

- A surplus of space and a shortage of tenants meant that rental levels had only one way to go: down. Nor was this a temporary hiccough. Rental levels were, on average, on a falling trend from mid-1990 to well into 1994.

- Strenuous efforts were made to disguise the extent of the downturn. Landlords began to offer longer and longer rent-free periods or other inducements to persuade a prospective tenant to sign a lease, probably at a rent above the open market rate. "Confidentiality clauses" became more common in new lettings, making it difficult to see what inducements had been offered and what true level of rent had been obtained.

- As rental levels moved down and growth prospects receeded, the most basic mechanism of the investment markets began to assert itself. Buyers of income-producing properties demanded higher initial returns to compensate them for reduced prospects of growth in the future. The only way that initial returns could rise was if prices fell. This is exactly what happened. Consequently, the values of virtually all types of property began to be marked down sharply.

• Another factor was at work. As we have seen, buyers of properties tend to compare the returns they are offered with returns on alternative forms of investment. If returns on alternative investments rise they will — all else being equal — expect higher returns from property. As interest rates rose and investors could get high returns from a simple bank account they came to look for higher returns from property. This again implied that prices must fall.

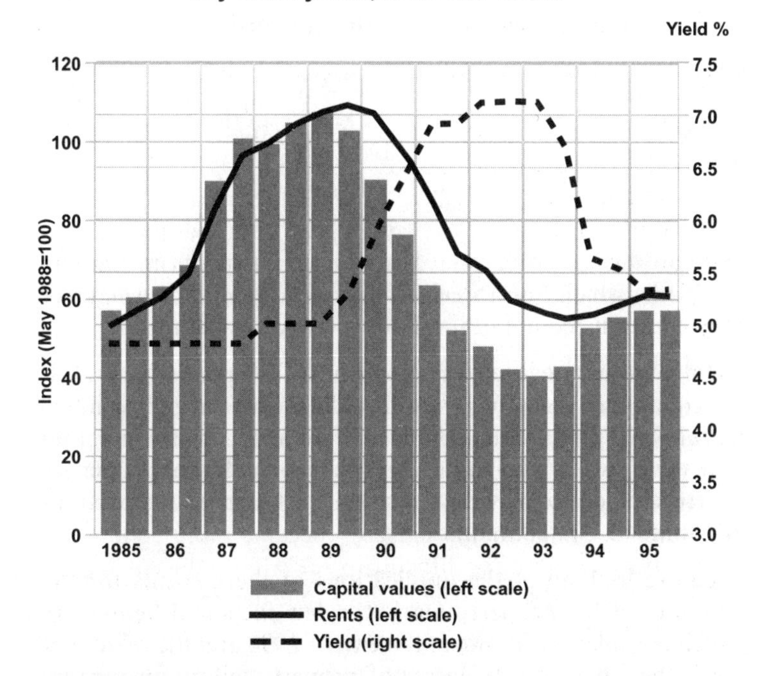

The mechanics of a market collapse: how falling rental values and rising yields combined to reduce the value of City of London offices by more than half in the recession. Source: *Hillier Parker*

• We therefore had at least three factors undermining property values. Since these values are basically a function of rental levels and yields, falling rents and rising yields exerted a multiplier effect on the fall in values. The fact that investors would have been looking for a higher return anyway to match the higher returns now available elsewhere was an exacerbating factor. The values of some City of London offices fell by more than half from their peak.

• The severe fall in values meant that many developers would have been bust even if they could have let their buildings and even if they could have found buyers for the resultant investments. The prices they would have received would have been too low to repay the bank loans. The fall in values also undermined the security for other types of bank loan secured on property. The major UK banks wrote off several billion pounds on dud property loans over this period.

• With would-be buyers in short supply, the number of transactions in the commercial property market contracted severely. This posed problems of valuation for surveyors deprived of adequate market evidence as a yardstick. Whereas much distress selling might have been expected, in practice it was not always easy to pick up good-quality revenue-producing property at bargain-basement prices. Owners (or banks which had foreclosed) tended to hold on wherever they could in the hope of a later recovery in values.

• Unlike some previous recessions in the UK property market, the collapse in values was world-wide. Thus foreign and international property groups often suffered as much as companies that were purely UK-based. And the ownership of properties overseas failed to provide a cushion for the more internationally-minded UK companies.

Before looking at the implications of these events for the structure of the property market, a few facts and figures are worth considering. Between the end of 1990 and the middle of 1993, the values of all classes of property fell on average by 38%. Worst hit was the office sector with values falling by about 53% on average over this period according to figures from agents Hillier Parker. City of London offices suffered appallingly, with average values down by 63%. Shops fared considerably better (or rather, less badly), with values falling on average by only 25% from their peak.

We stress again that these are average figures. Experience varied from one region to another and even within regions some properties fared better than others.

The falls in value resulted from a combination of rising yields and falling rental values. In the case of offices the fall in rental values was the most important factor. These had risen by 135% between November 1985 and November 1990 (and by 90% over roughly the same period in the case of City of London offices). They subsequently fell by some 43% from their peak levels, with City office rents falling by half. By comparison, the average fall in shop rental levels was comparatively modest: in the order of 9%. Across all types of business property, rents fell on average by about 28%.

Average yields for all types of business property rose to a peak of 9.4% in the first half of 1991 from 7.2% a couple of years earlier and remained close to that level for the next two years. It was not until the second half of 1993 that yields began to edge down again. And this was not caused by any improvement in rental growth prospects; rental values were still on a falling trend at that point. Capital values began to rise — and yields therefore began to fall — almost entirely because investors switched to property for a better return than they could get from bonds.

- The combination of falling rents and rising yields caught banking lenders on the hop. In their risk analysis of a property development they might have considered the effect on their calculations of yields rising by half a percentage point or so between planning and completion. As it turned out, yields often rose by three or four percentage points and values more than halved, leaving their loans seriously uncovered.

- The multiplier effect of all assumptions going wrong at the same time was dramatic both for developers and bankers. Suppose a developer did his sums in 1988. He planned an office building that would cost £100m including land and finance charges. He looked for a rent of £55 psf or £8.5m in total on completion and reckoned that the building would be valued on a 6.5% yield, or at 15.4 years' purchase of the rent. It would thus be worth around £131m. A profitable operation, with some room for contingencies. The banks agreed and were prepared to lend 80% of cost, or £80m. But by the time the building was completed in 1992 the rent the developer could hope to get had fallen to £30 psf or around £4.6m in total and the yield on which it would be valued had risen to 9% or 11.1 years' purchase of the rent. On this basis the building was worth only £51m, the developer was probably bust and the banks had lost a fair chunk of their £80m loans. In practice, the position was probably worse, since rising interest rates would have increased costs and it might have taken several years to find a tenant. We have taken a fairly extreme example, but greater or lesser horror stories lie behind many insolvencies among developers and severe bad debts among the banks.

• Because development land and developments in progress are generally valued on a residual basis (see Chapter 8), the values are highly geared. Any fall in the values of completed properties has a proportionately greater effect on the values of land or developments under way. This posed some difficult choices. Suppose a developer planned a scheme at an all-in cost of £40m which was expected to be worth £50m on completion. Most of the cost was borrowed from the banks. After he had spent £20m on the scheme, the developer went bust. The scheme passed to the lending banks. By this time, the likely value of the development on completion had fallen to £25m but it still needed a further £20m spent on it to bring it to this stage. The land and the work undertaken to date had virtually nil value to a prospective buyer — had there been one in the offing. Should the project be abandoned as it stood or should the extra £20m be spent? The sums and the answers varied, but in the mid-1990s you could see part-completed buildings in London and elsewhere on which no work had been done for several years. Land frequently acquired a negative value in the sense that the building you could put on it would be worth less than the construction and finance cost.

- The fall in rental levels destroyed one of the great assumptions of the post-war era: that rental levels would almost always be higher at the review point than they had been five years earlier. Now the property world was having to cope with the idea that there would, at best, be no growth in income at the review point.

- This introduced another phenomenon virtually unknown to many in the industry: the "over-rented" property. Responsibility lay with the "upward-only" rent-review clause in the typical 25-year institutional lease. This stipulated, in effect, that the rent would rise to a market level at each five-yearly review but never fall to a market level if this was lower than the passing rent at the time. Thus you could have a City of London office block with its rent set at £60 per square foot (psf) at the height of the boom in 1989. At the rent review in 1994 the market rental level might have come down to £35 psf or less. But the tenant had to continue to pay at least £60 psf until the end of the lease.

- Over-rented buildings introduced vast distortions into the market. The covenant of the tenant became all-important, ousting for a time the location and quality of the building as the main determinants of investment quality. There would be buyers for an over-rented building let to the government or a household-name company because the rent was secure. The same building let on the same terms to an insubstantial outfit would find buyers only at a much lower price, if at all. If the tenant was paying £60 psf but the market rate was only £35 psf, the landlord faced the prospect of having to re-let at £35 psf if the original tenant went bust. Likewise, over-rented properties with only a short lease period to run were not regarded as a great investment.

- The over-renting phenomenon also threw into focus the "privity of contract" principle in English law. In effect, this meant that the original lessee remained ultimately responsible for the rent, even where he had assigned the lease with the landlord's consent. It did not seem to matter too much when rents were rising. If the subsequent tenant defaulted, the property could be easily re-let at a rent as high or higher. With falling rents, the principle could give rise to a serious liability. There were harrowing stories of small shopkeepers who had sold their business and their lease on their retirement many years ago when the rent was £10,000 a year. The rent had subsequently risen in stages to, say, £40,000 a year, then the current holder of the lease defaulted. The original shopkeeper then found himself responsible for a £40,000 rent on a building that would fetch a much lower level in the market after the events of 1990 to 1994. Even some of Britain's major retailers had to provide for substantial financial liabilities incurred in this way. The authorities eventually decided to do away with the privity of contract principle in respect of new leases, with effect from 1996. But even in the case of leases originally signed after the beginning of 1996, tenants who wished to assign their lease subsequently were generally required by the landlord to provide a guarantee against default by the immediate assignee.

- Over-rented buildings also compounded the problems of the valuation profession during the recession years. Not only did the valuer have to pay far more attention than normal to the covenant of the tenant. He also had to take account of vastly different views in the market as to the value of such buildings. The "spread" between prices that sellers were prepared to accept and buyers were prepared to pay widened enormously, often ruling out the possibility of a deal. This was at a time when valuation posed severe problems anyway because of the lack of transactions in the market to act as a yardstick. We look at some of the problems in Chapters 47 and 49. In 1994 the Mallinson committee on property valuation emphasised that "open market value" had to assume that an owner was prepared to sell at the best price a buyer would have been prepared to pay after an appropriate marketing period — even if a prospective seller might in practice have decided not to sell at this price.

- Concerns about valuation were not confined to the property industry. Industrial and commercial companies complained that they were having to write down the value of the properties they owned in their accounts, thus weakening their balance sheets. Some argued that this was illogical, since the buildings had not become any less valuable for use in the business. Why should they have to write down these values simply because there had been a collapse of confidence in the highly volatile property investment market — which was no concern of theirs? Some also questioned whether valuers could produce reliable figures with so few transactions to take as a guide. Perhaps some other method of incorporating property values in the accounts would be more appropriate? Most of the complainants conveniently ignored the fact that they had been only too happy to write up the values of their properties in the late 1980s, when this same derided property investment market was booming. By 1996 the Accounting Standards Board was proposing to stamp on this "cherry-picking" approach to valuation. Companies could choose to carry the properties they used in their business at cost or at revalued figures. But if they plumped for revalued figures, they must revalue regularly — not just when it suited them.

- "Reverse premiums" became commonplace as an accompaniment of the over-renting phenomenon. Suppose a business owned the lease of a London City office block that had become surplus to its requirements. Suppose, too, that the rent had last been set at £60 psf and the market rate would now be £35 psf. To assign the lease at £60 psf to another tenant, the original tenant would have to offer a large inducement. This probably took the form of a lump-sum payment to compensate for the element of above-market rent the new tenant would have to pay. A lease on an over-rented building could thus be a significant liability, and the accounting bodies attempted to produce rules on how these liabilities for the landlord should be dealt with (see Chapter 16). The problem with reverse premiums was that, even after forking out a large sum to get the lease off its hands, under the privity of contract principle the original tenant would still be liable for the rent if the new tenant defaulted.

• As we have seen, premiums were also paid by developers and other property owners to persuade prospective tenants to sign a lease. These might take one of several forms: a long rent-free period or a lump sum payment (perhaps described as a "fitting-out allowance") were commonplace. Whereas the tenant might enjoy up to six months rent-free even in normal times, during the property recession the period often extended up to three years or so. This posed problems in working out the true rental value of the building. If the tenant signed up at £40 psf and received three years rent-free, it was a reasonable assumption that the building would fetch a rent considerably below £40 psf without the inducement. We look at this problem in Chapter 48. Again, it posed problems for valuers trying to deduce rent comparables from transactions involving large inducements.

• There was another aspect to the "inducements" problem. Sometimes the developer, anxious to attract a business to his new building, would offer to take over responsibility for the lease on the business's existing premises. If the building was severely over-rented, this could be a hefty liability for the developer. If the prospective tenant owned his existing building, the developer might buy it at an above-market price. There was, of course, a severe problem for the business if the developer subsequently went bust and defaulted on the lease he had assumed.

• With an over-supply of space and tenants in a strong position relative to the landlord, the traditional 25-year "institutional" lease with five-yearly upwards-only rent reviews began to crumble. The events of 1990 to 1993 made prospective tenants very much aware of the liabilities they might be assuming. Frequently they bargained for shorter leases, for leases with break clauses or for leases with special protection built into the rent-review clauses. Many landlords, however, preferred to offer large lump sums or rent-free concessions rather than compromise too far on the structure of the lease.

• The structure and length of the lease became an increasingly important factor in determining investment values. The gap widened sharply between values of buildings let on traditional 25-year leases and those let on shorter leases or with break clauses (see Chapter 51). From the financing point of view, leases that the tenant could break within ten years found little favour with banking lenders (which was one reason for landlords' reluctance to make too many concessions on this score).

• Bank lending to property companies more than doubled to £34bn between 1988 and 1990, having risen about sevenfold since 1985. It eventually peaked at over £40 billion in 1991 and began to run down slowly thereafter. Bankers were desperately anxious to reduce their property exposure and it became difficult to raise fresh loans against property and virtually impossible to finance new speculative development. By the middle of the decade the position had eased a great deal, at least in relation to revenue-producing investment property with a long lease and good covenant. The banks were competing strongly for this type of business. But finance for speculative development was still frequently unobtainable and many post-war financing assumptions had been shaken.

• Since the late 1950s the rental yields on investment-grade property had generally been below borrowing costs: often a long way below. This "reverse yield gap" posed the central property financing problem of the post-war years: how to cover interest charges on loans taken out to develop or buy property. Many different methods were used. In essence most of them boiled down to techniques for conceding future growth in return for cheaper finance at the outset. They ranged from leaseback development finance to stepped-interest or zero-coupon loans. The assumption was that rents and therefore values would rise and that the total return — comprising rental income and growth in capital value — would exceed borrowing costs. If rents did not cover borrowing costs at the outset, they would rise to do so at the first review. But the events of 1990-93 knocked these assumptions on the head. Rental levels fell and capital values fell. There was — for the time at least — no growth to swap for cheap finance. Existing borrowing arrangements based on assumptions of future growth generally had to be refinanced in one way or another.

• At the same time as it became difficult to raise money on terms that assumed future growth, the reverse yield gap itself was eroding. With average yields on investment-grade property rising above 9% and government bond yields coming down below 7% at one point, in theory it became possible to finance some types of property with loans and cover the interest with the rental income. At the very least, higher property yields and lower borrowing costs considerably reduced the financing gap.

• Buyers of over-rented buildings recognised that they were in reality buying two different flows of income. There was a "core" rent, which was what the building would produce if available for a new letting at the time. And there was the "froth" element consisting of the rent that the tenant was paying above the market rate. The froth was not a genuine growth income stream. It had more in common with interest payments on a loan. It followed that property values — and particularly values for over-rented property — came to reflect more closely the returns available to the investor from bonds and other kinds of loan. Thus, when property values began to recover in 1993 and 1994, it was at first mainly because bond yields had come down, not because perceptions of future growth had improved greatly. Since the income returns that investors could obtain from bonds had dropped heavily, they looked for other types of investment offering a better income. Property — particularly over-rented property — at bombed-out values was the obvious candidate, and the buying by investors forced prices up. The other side of the coin was that values began to look vulnerable again when bond yields moved back up in 1994.

• The central question during the property market recession was whether property could ever fully regain the growth status it had enjoyed for most of the post-war period. By the mid-1990s the answer was still far from clear. It was assumed that the surplus new space — particularly in City of London offices — would eventually find tenants, and this was happening. The problem lay more with the older or less attractive properties from which tenants had moved out and which might never be re-occupied in their existing form. But the new space had to be mopped up and rents had to begin rising again before it would become economic to redevelop the older buildings. At best, and in a more relaxed planning environment, it seemed unwise to assume the shortages of good-quality property that had fuelled rental growth for much of the post-war period. And if inflation was to remain at a permanently lower level, property values could not look for a boost from this source.

So much for the effects of the crash on the property industry. There was also hot debate — though more from outside the property industry than from within it — on the effects of the commercial property crash on the economy as a whole. Critics argued that upwards-only rent review clauses in leases had slowed the progress of economic recovery. This was because they prevented rents from dropping back to market levels and thereby reducing business occupiers' costs and encouraging them to trade out of the recession. And the secrecy surrounding the terms of new lettings and the inducements offered had resulted in a distorted and opaque market in commercial property.

The property industry and property investors by and large argued that "the market" would sort out problems of this kind. They pointed to the changes in the terms of commercial leases resulting from the greater bargaining power of prospective tenants in the recession. And they argued that long leases and upward-only reviews were necessary to provide security to investors and encourage them to provide the volumes of finance that the industry needed.

It was noticeable, however, that the "market will take care of it" school of opinion tended to ignore the cyclical nature of the property industry and ignore the fact that many property markets overseas operate satisfactorily with shorter leases and an absence of "upward-only" clauses. This cyclical factor ensures that measures taken in any one climate will have their effect in a very different one. The market addresses yesterday's problems rather than tomorrow's. Thus, when the property industry turns up, landlords will doubtless again be able to impose longer leases and upwards-only rent reviews. The distorting effects on economic activity would not emerge until the next cyclical downturn.

The government decided to look into the matter and asked for views. After a lengthy study of the "upward-only" question and various other property market distortions thrown up by the slump, it decided in the summer of 1994 to take no action on the structure or transparency of commercial property leases. Sheltering behind the traditional fig leaf of a code of conduct to be devised by the industry, the government, too, decided that the market could sort it out.

In the following chapters we look in more detail at some individual aspects of the property market slump. These chapters are again based on articles that appeared in *Estates Gazette* while the events they describe were unfurling. Techniques for evaluating rent-free periods and for valuing over-rented properties were constantly evolving during the recession, and the theoretical approaches sometimes seemed to be at variance with such evidence as the market provided. We do not therefore claim for a moment that these chapters provide the last word on the subject — if, indeed, there is a last word. They serve more to identify the problems and describe some of the techniques evolved at a particular point in an attempt to solve them.

47 Valuing over-rented properties

In the property market slump of the early 1990s many buildings in Britain, particularly offices in the City of London, became "over-rented". In other words the rent which the tenant paid was above the current rental value of the building: the rent it would have achieved if let in the open market at the time. The problem grew more acute as the property market's problems deepened and rental values in many areas tumbled. It was a problem with which valuers had rarely had to contend since the property collapse of the mid-1970s, and it surfaced in the early 1990s in considerably more acute form and looked set to last considerably longer.

Attitudes to valuing over-rented buildings changed and evolved as the property market slump progressed. Even by the mid-1990s when the worst of the decline seemed to be past, it would have taken a brave man or woman to maintain that there was one single correct and universally accepted method of putting a value on a building let at a rent above the going market rate.

What follows in this chapter is based on an article that appeared in *Estates Gazette* in the autumn of 1991, right in the middle of the market slump. The examples and interest rates taken are therefore of that period. It is certainly not the last word on valuation of over-rented property. Its interest is more a historical one, reflecting the perception of the problem and one strand of thinking on the solutions at the time. As thinking developed there was a greater tendency to treat the rental flows from over-rented property as a form of income derivative — subject to totally different pricing techniques — rather than to attempt to rationalise them in traditional property terms.

So how, in the middle of the 1990s market slump, would you have approached an over-rented property? The starting point would probably have been in line with the principles set out in the early chapters of this book. But before getting into the detail, one important general point must be made.

The most common form of lease for properties of investment quality before the recession struck was the so-called "institutional lease". Typically the lease ran for 25 years with "upward-only" rent reviews every five years, and the building was let on full repairing and insuring terms (the tenant was responsible for these items). The important point for our purposes is the upward-only rent reviews every five years. They mean that, even when the rent review becomes due, the rent which the tenant pays will not drop back to market level. If the building were originally let at £60 per square foot (psf) and would fetch £40 psf today, the tenant continues paying at least £60 psf for the duration of his lease. (This is on the assumption that the tenant does not go bust in the interim — an important consideration to which we will return later.)

So let us suppose that a 10,000-sq ft City of London office block had its rent reviewed to £60 psf at the height of the rental boom: say, just before mid-1989. The rental income from the building was £600,000 a year, but if it had come on the market as a vacant property in 1991 the landlord would have achieved a rent of only £400,000. Let us suppose, too, that in 1991 the lease had eight years to run before it expired; there would have been one more rent review in 1994 and, provided the tenant stayed solvent, the rent could not have been reduced below its existing £600,000 a year level because of the "upwards only" clause.

Thus the landlord was, in 1991, sure of getting at least £600,000 a year for the next eight years, provided that his tenant did not go bust. If this £600,000 had represented the current rental value of the building, and if a building of this type would have been valued on a yield of, say, 7% or at 14.3 years' purchase of the rent, it would have been worth about £8.6m in 1991 (14.3 times £600,000).

Unfortunately, in practice it would have been worth considerably less. Look at what the owner was really getting. He received £400,000 a year which represented the real rental value of the building. And on top he received £200,000 of "froth", which was the amount by which the building was over-rented as a result of the historical accident of its having being let at the peak of the rental market.

The £200,000 of froth clearly could not be valued in the same way as the £400,000 of core rental income. The owner had to rationalise the situation. He was getting £400,000 a year which he would have been pretty sure of getting even if the tenant went bust and he had to find another tenant. This £400,000 a year was therefore the base from which any estimates of future income growth would have been calculated.

In addition he was getting the £200,000 a year of froth. This was clearly worth having, but it was income of a much lower quality and he was sure of getting it only for the remaining eight years of the lease (and only provided that the tenant stayed solvent). After eight years, when the lease expired, either the tenant moved out or a new lease would be negotiated at the then current market rent. The element of over-renting would have disappeared.

So if the building was one which would have been valued on a 7% yield or at 14.3 years' purchase if let at a current market rent, we would have had to apply the years' purchase to the £400,000 of rent. This would have given a value of £5.72m for this part of the sum.

But the £200,000 of froth would also have had a value, albeit a much lower one. Since it was income which was not going to grow we had to value it at a market interest rate rather than at the 7% yield which had growth assumptions implicit in it. Suppose we valued the £200,000 on an 11%

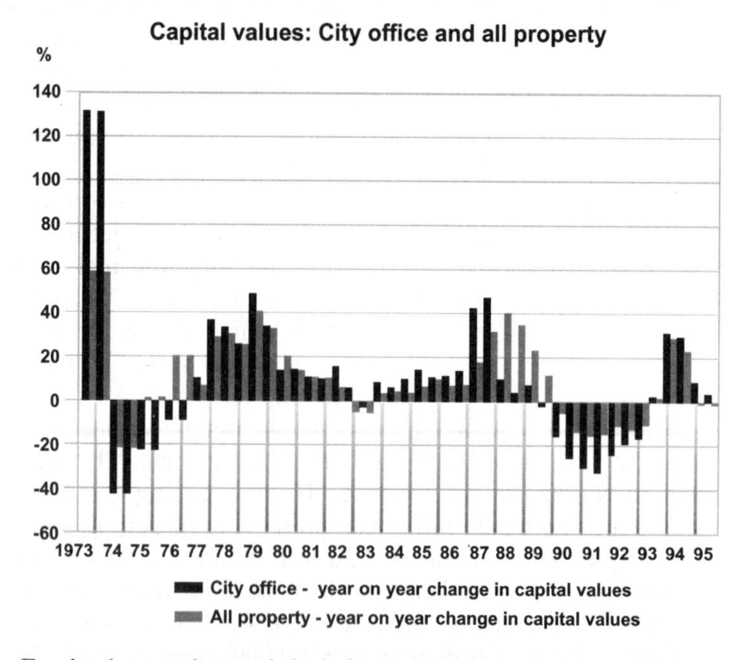

For developers who get their timing right there is money to be made in City of London offices. For those who get it wrong and produce the new space as demand is turning down, disaster awaits. Our chart illustrates the boom-and-bust cycle in City office values and shows that the City office cycles may peak and bottom out at different times from the cycle for property as a whole. The peaks and troughs are also generally more pronounced. Source: *Hillier Parker; Datastream*

yield. We would also have needed to allow for the fact that the landlord would have received the "froth" for only eight years.

At this point there was a difference of opinion. Some valuers would have argued that the £200,000 should be valued on a simple years' purchase, without any allowance for a sinking fund. Others would have argued that, since it would disappear after eight years, it was much like a short-leasehold investment and needed to be valued with allowance for a sinking fund. The arguments on both sides were complex, but somewhat beyond our scope here. For simplicity we have adopted the simple years' purchase (which is not, however, to come down on one side of the argument or the other).

Valuation tables show that, taking an 11% interest rate, the right to receive £1 per year for eight years is worth £5.1461. So the right to receive £200,000 a year for eight years is worth £1.03m (using a leasehold valuation with sinking fund would give a lower figure). On this basis, our property would have been worth £6.5m (£5.72m plus £1.03m). This was a far cry from the £8.6m which it would have been worth if the whole of the £600.000 rent represented the rental levels of 1991. However, a buyer of the property would actually have been receiving £600,000 a year, so the running yield if he bought the building at the valuation figure would have been around 8.9% on his £6.75m outlay. On the face of it, and without going through the calculations we have just examined, this would have looked very high.

And this is perhaps one of the reasons why the market got itself in such a muddle about over-rented buildings during the early 1990s property slump. In the first place, that 8.9% running yield depended entirely on the tenant's ability to stay solvent and continue paying the rent for eight years. If he went bust and the building had to be relet at £400,000 a year, the running yield on the £6.5m outlay would have dropped to 5.9%. So quality of covenant was all important. If the tenant was a weak one and stood a fair chance of going bust within the eight years, you would have taken a far higher rate of interest than 11% to value the £200,000 of froth income. In other words. you would have put a much lower value on it and therefore on the property as a whole.

The second problem is rather more complex. We have been talking so far about conventional valuation. But there was evidence from the market in 1991 that at least some buyers did not look at over-rented properties in this way. They were more inclined to undertake a discounted cash flow evaluation of the investment, which involved making specific assumptions about rental flows each year and — very important — about the yield on which the building would be valued at the end of the eight years.

The important point was that some buyers, particularly foreign ones, appeared at the time to be arriving at a higher value for the property using these methods than they would have done with a conventional valuation. In theory, if the assumptions which were spelled out or implicit in the two approaches were the same, they should have given much the same result. In practice, perhaps the discounted cash-flow supporters were taking a more optimistic view of future yields or rental growth, were making less allowance for the risk inherent in an over-rented building or were simply satisfied with an internal rate of return rather lower than that for which many British buyers were looking.

Whatever the answer, there was evidence of a dual standard of values during the property slump which greatly complicated any dealings in over-rented properties for a considerable time.

48
Phoney rents and true rents

A landlord lets a building to a tenant on a standard 25-year institutional lease, but allows him the first three years rent-free as an inducement to sign. The quoted rent is £35 per square foot (psf). What is the true rental level for the building?

Problems such as these were common in the commercial property market collapse of the early 1990s particularly in the case of City of London offices where landlords were prepared to offer big inducements to secure a tenant. At least five forms of inducement were common:

- A long rent-free period at the start of the lease.
- A "capital contribution". The landlord offered the tenant a capital sum as an inducement to move in.
- A "take-back" — the landlord either bought the prospective tenant's existing property or took responsibility for his existing lease.
- A "stepped rent". The rent started at nil or at a low level, rising by predetermined steps in the earlier years of the lease.
- A "capped rent". The landlord agreed that the rent would not rise more than a certain amount at the first review.

In practice, no two deals were exactly the same and they might incorporate a mix of two or more of these inducements. But all posed the same problem: how did you calculate the true rental level for the building? This could be important if the letting in question were to be cited as an example when attempting to establish the market rent of a comparable nearby building at the review date. Again, this chapter is based on an article that appeared in *Estates Gazette* in 1992 and reflects the thinking of the time.

Taking our original example of the building let at £35 psf with three years rent-free, there were a number of more or less simplistic ways in which the case could be argued:

View number one: "There is no rent for three years, then £35 psf is payable for the next two years up to the first review. The average rent for the first five years is therefore £35 multiplied by two and divided by the five years: in other words, £14 psf."

View number two: "Nonsense. The three years rent-free was granted as an inducement to take a lease for 25 years, not five. So the benefit has to be spread over the full term of the lease. The tenant is saving £35 psf for three years, which comes to a total saving of £105 psf. Spread this over 25 years and it amounts to an annual saving of £4.20 psf. So the true starting rent is £30.80."

Both these arguments were subject to a considerable number of objections. An overriding problem was that neither took account of the time value of money. Because the tenant notionally earned interest on the money he was not having to pay as rent in years one, two and three, the £35 that he saved in year one was worth more to him than the £35 he saved in year two. And the £35 saved in year two was worth more than the £35 saved in year three.

A similar problem arose if, instead of offering a rent-free period, the landlord offered a capital sum as an inducement for the tenant to move in. Suppose he were to offer a lump sum equivalent to three years' rent: in other word a sum equivalent to £105 psf, payable at the outset. This was clearly worth more to the tenant than a saving of £35 a year for three years, because the tenant could earn interest on the full amount of £105 from day one.

The only realistic way of analysing the effect of various forms of inducement or subsidy was a discounted cash flow (DCF) calculation, which could take into account the form and timing of the inducements. This was not to say, however, that the market would always adopt such an approach, so theory and practice sometimes diverged.

The point about a DCF calculation was that it forced you to spell out your assumptions. Probably the most important of these was the "crossover" point at which the true rental value of the building would rise to overtake the "headline" rent initially charged. In our example, this would have been the point at which the true rental value of the building would rise above £35 psf. You might have assumed that the true rental value would be above £35 psf by the time of the first review at year five. You might have assumed that this would not happen until year 10 or later.

For convenience we have taken a building let at a rent of £lm a year, which equated, say, with a rent of £35 psf. Assume that, even in "normal" times in the property market, an incoming tenant might have expected a three- to six-month rent-free fitting-out period. We have taken six months for simplicity.

There are still a number of different approaches that we could have taken in the calculation. But the important point was to decide at what stage you expected the rental value of the building to grow to exceed the headline rent (£35 psf or £lm in total, in our example). This was the "crossover point". If you decided that it was not likely to be before year 10, in effect the tenant — if he took the three-year rent-free period — would have been paying nothing for three years, then an over-the-odds £lm a year rent in years four and five. Since we have assumed that the true rental value of the building would not have achieved £lm by the end of year five, he would have continued to pay the over-the-odds £1m a year rent in years six to 10. At this point we assumed that the true rental value would have reached £lm.

Looked at this way, in return for the rent-free period, the tenant would have been paying an over-the-odds rent for the following seven years. Suppose, instead, he had opted to pay a market rent from the outset (or, rather, from the end of the customary first six-month rent-free fitting-out period)? What should that rent have been?

We could have found out by seeing what rental figure would have cost the tenant the same amount over the 10 years — allowing for the time value of money — as his three years rent-free and £lm a year thereafter. Suppose we had taken a discount rate of 12% as representing the cost of money. First, we would have discounted back to a present value all the £lm a year rent payments (or, more accurately, £250,000 per quarter) that the tenant would have paid in years four to 10. Totalling these present values together would have given the answer of some £3.49m. This was the present value of the rent paid over the first ten years of the lease. What rent (payable from the end of the first six months), if discounted back in the same way, would have given the same answer?

There was a complication here in that we had to assume a growth rate in the true rental value of the building. This meant that the rent which the tenant paid — if he started with a true market rent — would rise after year five, even though it would still be below £1m a year. So a little trial-and-error calculation was needed.

In practice, the calculations showed that, at a 12% discount rate, a starting rent of about £546,000 a year would have given the same present value of £3.49m as a rent of £1m a year with three years rent-free. What growth rate in the true rental value did this assume? Around 6.5% a year compound, which meant that the rent paid increased from £546,000 a year to £748,000 at the five-year review and the rental value just passed £lm at the end of year 10. This sort of growth rate would have been within the range of reasonable assumptions. And the £546,000 starting rent would have equated to £19.08 psf in our example.

As a check, we could have undertaken the same calculation assuming a crossover point at the end of the first five years. In this case the true rental value would have been only £382,000 a year or equivalent to £13.37 psf. And this value would have had to grow at over 22% a year to reach £lm by the end of year five. As this looked clearly unrealistic, the assumption of a crossover point at year 10 was probably the correct one. Thus — given our assumptions — a rent of £19.08 psf with six months rent-free would have cost the tenant the same, and given the landlord the same return, as a rent of £35 psf with a three-year rent-free period. That was the theory. Then it came down to the haggling.

49
How rent-free periods affect values

The commercial property market collapse of the early 1990s threw up many new and unfamiliar problems for valuers. Overrented buildings were by no means the only one. Properties with long rent-free periods and the like could be evaluated in a number of different ways, and there might be wide discrepancies between the expectations of would-be buyers and would-be sellers.

One result was a growing emphasis on the use of discounted cash flow (DCF) techniques to evaluate unconventional rent flows. Our calculations here are based on the situation as it appeared in 1992. Again, techniques were constantly being developed and amended during the course of the slump.

How, as a starting point, would you have evaluated the Daily Telegraph's lease of part of the Canary Wharf Tower in London's Docklands? Remember that this immense office development, constructed by the Canadian developer Olympia & York, hit the market just as the surplus of office space in the nearby City of London itself was leading to empty buildings and plummetting rental levels. Canary Wharf stood largely empty for several years despite massive incentives for tenants to take space, was a major factor in bringing Olympia & York to the ground, and effectively passed into the ownership of the banks before its prospects began to improve as the slump drew to its close.

According to the Telegraph's prospectus when it floated on the Stock Exchange, the Telegraph was paid an "inducement" — its own term — of £20m to move in to the tower, as well as having its existing Docklands building purchased for a further £20m — considerably more than was likely to have been worth at the time. In return, the Telegraph agreed a lease at about £30 per square foot (psf), whereas the going rent in 1992 (if there had been such a thing) might have been closer to £12-£15 psf. Difficult, indeed, to value this income stream in terms of an all-risks yield, since it had little to do with property comparables and everthing to do with evaluation of risk and of the value of a corporate cash flow.

In many situations, valuers would try both a traditional all-risks yield approach and a DCF calculation, and re-examine their assumptions if the two threw up vastly different answers. But, where there were few real-life transactions as a guide, even this may have been a relatively poor indication to what would actually happen in the market-place.

Take a building where the tenant was granted a long rent-free period. How would you have valued a building that would not produce any rent for a couple of years or more? The quick answer was that no developer in his right mind would have sold the building until it was producing an income flow. But, if there was a bank breathing down his neck, he may not have had the choice.

The second answer is that there were no theoretical problems in putting a value on a deferred income flow. You could adopt either the traditional all-risks-yield approach or the DCF one. Assume, for simplicity, that we are talking about a new freehold office block let to a good tenant on a 25-year lease with five-yearly upward-only rent reviews. And that, if it were producing rent immediately, a valuer might have reckoned that a purchaser would have been prepared to buy it at a price which showed him an initial yield of 9%. In other words, it would have been worth 11.1 years' purchase of the rent, or £1.11m at a rent of £100,000 a year.

Then suppose, taking an extreme example to illustrate the principle, that to sign up his tenant the developer of the office building was obliged to offer a three-year rent-free period. What was the building worth, with a tenant signed up but with no income forthcoming for three years?

In theory, the answer should be reasonably simple, if you assume that £100,000 a year was a fair current rent for the building and there was thus no element of overrenting. In three years' time, when it began to produce rent, the building would have been worth £1.11m on the assumptions of the period. So you would have discounted the £1.11m back for three years to arrive at a present value, using as your discount rate the 9% all-risks yield. The answer would have been £858,000. In other words, £858,000 was the figure which, at 9% interest, would have compounded up to £1.11m in three years. So £858,000 should have been the present value.

You could then have run a DCF calculation as a check. Suppose you assumed that, given the oversupply of office space in London at the time, there would be no growth in rental value for the first five years of the lease. Thereafter, between years five and 20 you might have assumed that the rate of office rental growth would have returned to a little below its long-term post-war average — say, 5% a year. So in year 11 the rent would have risen to £127,628, in year 16 to £162,890 and in year 21 to £207,893. By then the building would have been ageing, so perhaps you would have assumed that the rate of growth tailed off to 2% a year for the last five years of the lease.

What would the building be worth at the end of the lease? You could have made the assumption that it would be in need of extensive refurbishment or even redevelopment and have calculated a residual site value. Or, for simplicity, you could have assumed that it would be valued on a considerably higher yield than that of the day to reflect its age: say, 12%, or 8.3 years' purchase. On this basis it would have an end-value of £1.9m.

You would have tried the calculation, first, on the assumption that the rent flowed from year one. Discount back each year's rent for the relevant period — one year for the first year's rent, 25 years for the final payment — and discount back the end-value of £1.9m for 25 years. By trial and error you would have found that a discount rate of around 11% the sum of all these discounted values came to about £1.11m.

In other words, if you had paid £1.11m for the building and your assumptions had been correct, you would have earned an internal rate of return or IRR of 11% on your money. If you knew that at the time you could have achieved a totally safe return of just under 9% on a long-dated government bond, this perhaps offered a reasonable margin for risk and the relative illiquidity of the property investment.

But this was on the assumption that the rent flowed from day one. In practice, because of the rent-free period, you would have been receiving no rent in the first three years. So you could then have performed the same calculation, but leaving out the rent for the first three years. If you were aiming for the same rate of return of about 11%, this would show that the building would have had a present value of £872,000 (the total value of all the rental payments and the property's assumed end-value, discounted for the appropriate periods at an 11% discount rate, would have been £872,000). This was marginally higher than the answer you came up with via the other method. (In practice, the DCF calculation might have been performed over a shorter period of time, rather than the full term of the lease. We have taken the full-term calculation to demonstrate the principle).

The discounted cash flow approach was fine in terms of calculating what the building was worth to you. Unfortunately, it was a somewhat theoretical exercise if there were few, if any, buyers able or prepared to buy a building that was not producing income.

So, in practice, if the property had to be sold, what might have happened was that an "artificial rent" would have been created. This could be done by putting a portion of the sale proceeds into an "escrow" account, out of which the purchaser was paid a sum equivalent to the rent on the days when the rent would have fallen due. The purchaser might have argued: "I'm losing £100,000 a year for three years, so £300,000 should be knocked off the £1.11m purchase price and put in an escrow account to pay me my £100,000 a year missing rent." This would have meant that the vendor would have received only £811,000 for his building.

The vendor would have argued that the money in the escrow account would earn interest, and it would therefore require appreciably less than £300,000 to make up the missing rent payments on the due dates. But, if a rate of interest was to be assumed, what should that rate have been?

One solution sometimes adopted was to agree the proportion of the sales proceeds which should be withheld from the vendor and paid into the escrow account. If, because interest rates turned out higher than had been assumed, there was something left over in the escrow account after making good the missing rent payments over three years, this residual amount reverted to the original vendor.

These and similar arrangements, which may well not have been publicised at the time, added considerably to the problems of deducing reliable market data from such transactions as took place in an illiquid market.

50
Why landlords like headline rents

Why are landlords so reluctant to drop their asking rents, even at a time of crisis in the commercial property market when prospective tenants are few and far between? In the early 1990s, office developers in London were prepared to concede all manner of inducements to persuade potential tenants to move in: long rent-free periods, large lump sums as "fitting-out allowances". Would it not have been simpler to drop the asking rent?

It might have been simpler, but it would not necessarily have been more commercially sensible. The structure of property financing and of the typical "institutional lease" has introduced many distortions into the commercial property market in Britain. Landlords have often had an incentive to press for a high — even artificially high — rent.

If we go back 20 or 30 years, this was largely a function of the financing pattern. Take an office developer who, when his project was completed and let, planned to repay his construction finance from the proceeds of a long-term loan against the property. He would probably have expected to raise a mortgage of two-thirds of valuation.

Suppose he had planned his property to produce a rent of £100,000 a year. Valued on a 6% yield, say, or at 16.7 years' purchase of the rent, it would have been worth £1.67m. A mortgage of two-thirds of value would have produced £1.11m.

But what if he held out for a rent of £120,000? If he achieved it, his building would have been worth just over £2m and a mortgage of two-thirds of value would have brought in £1.34m. Even if he had needed to sit on the empty building for a year until he achieved the higher rent, it would have been worth it. He would have foregone £100,000 of rent, which would have been subject to tax, anyway. In return he would have brought in more than £200,000 of additional cash.

Sums like these were supposed to account for many empty office buildings in London at the time. Harry Hyams, the legendary developer of the Centre Point office block on London's Tottenham Court Road, was reputed to increase his asking rent whenever a tenant looked like taking the building — a suggestion he always strenuously denied.

Fact or fable, games such as these would have been possible only at a time of office shortage and rapidly rising rents. The position in the early 1990s was very different. The City of London was awash with empty offices and prospective tenants were in very short supply. There was a surplus of shopping space in many areas. So why were landlords so reluctant to cut rents by the additional amount necessary to secure a tenant?

Much of the answer lies in the "upward-only" review clauses built into the lease. The traditional "institutional lease" (a lease of the type which the investing institutions like to see on the properties they buy) runs for 25 years with rent reviews every five years. At the review point the rent may increase, but cannot fall. This lease pattern had been breaking down somewhat in the depressed property market conditions of the early 1990s, but the "upward-only" clause was vigorously defended.

So a landlord who let his building at, say, £35 per square foot (psf) under the traditional lease structure knew that he was getting at least £35 psf for 25 years — provided his tenant remained solvent. Any buyer of tenanted property as an investment knew the same. Suppose the building in question was of 100,000 sq ft and would have been valued on an 8% yield in a depressed market, or at 12.5 years' purchase of the rent. The total rent would have been £3.5m. Provided the valuer was prepared to accept this as the market rent, the building could be worth £43.75m. If the landlord had had to offer the tenant two years rent-free at the outset to attract him in, the building might have been worth £43.75m less the discounted value of the two years' rent which the landlord had foregone: say about £6.5m to give a value of £37.25m.

If, on the other hand, the developer offered no rent-free period but agreed to accept a rent of £25 psf, the total rent roll would have been £2.5m and the building might have been valued at only £31.25m if £25 psf was thought to be the market level of rent.

The basic principle was that each £1 spent on offering the incoming tenant a rent-free period or a lump sum "fitting-out allowance" cost the landlord a maximum of £1 (ignoring tax factors for the moment). But each £1 conceded as a reduction in the asking rent cost the landlord (in our example) a £12.50 reduction in the capital value of his building.

The cost might, indeed, have been even higher than this. Take a parade of 10 identical shops, all belonging to the same landlord and all let on leases with upward-only rent review clauses. A rent of £10,000 a year each was established when the rents were last reviewed four and a half years earlier. In the meantime the rental value rose sharply (to £18,000 at the peak, the landlord calculated), then slipped back in the recession. But he still hoped to be able to achieve a rental increase to £14,000 for each shop when the rent was reviewed in six months' time. If he succeeded, valued on an 8% yield or at 12.5 years' purchase of the rent, each shop would have been worth £175,000 at that point, giving a value of £1.75m for the whole parade.

But a tenant went bust and the landlord found himself with one empty shop which was currently costing him £10,000 a year in lost rent. He tried unsuccessfully to find a new tenant at £14,000. In the recessionary climate there were prospective tenants who would take the shop at £10,000 a year. But rather than let it at this level the landlord kept the shop empty.

The point is that, if he conceded a letting at £10,000, it would have cost him dearly when the rents on the remaining shops came up for review in six months. The other nine tenants would have argued that, on the precedent of the latest letting, £10,000 was the going market rate for shops in the area. And they would have contested bitterly his demand for an increase to £14,000.

If they won and their rents were fixed again at £10,000 for the next five years, each shop might have been valued at £125,000, giving a value of £1.25m for the whole parade. So, by conceding a reduction in the asking rent on one shop, the landlord would have reduced the value of his total investment from £1.74m to £1.25m. It was cheaper for him to keep several shops empty than to make this sort of concession.

The example is simplified and is, perhaps, an extreme case. It also ignores any tax considerations that entered into the landlord's thinking. But it helps to explain why empty shops were such a common feature of many High Streets during the 1990s, and why rents failed fully to adjust at the margin to changing economic circumstances.

The position was exacerbated by several factors. Insurance firms and pension funds are important owners of much High Street property and, since they are investing with their own (or their savers') money, they do not have to worry too much about the loss of cash income from a few empty shops. The long-term capital value of their investment is far more important to them.

A landlord whose shops are financed with borrowed money is in a rather different position because he needs the cash from the rents to meet his interest charges. Loss of income from one or two shops could cause him problems. On the other hand, conceding a reduction in the asking rent that severely dents the capital value of his investment will decrease his borrowing power. And, if he is highly borrowed, it might, in an extreme case, reduce the value of his assets below that of the money he owes — in other words, he could be insolvent.

There are thus many pressures in the British property market which prevent rents adjusting rapidly to a change in the economic climate. The length and severity of the recession of the early 1990s did, it is true, see asking rents for some classes of property fall heavily from their peak levels. Had the structure of the market allowed this process to happen a little more rapidly and go a little further, rents would perhaps more quickly have established a base from which they could have begun to climb again. And property would more quickly have regained its status as a growth investment.

51 Valuing with shorter leases

Some of the most basic assumptions on which the commercial property market was built were thrown into question by the slump of the early 1990s. In particular, the structure of the typical "institutional" lease came under attack. This chapter, based on articles published in *Estates Gazette* in 1992, reflects some of the market thinking of that time.

To see what happened, and why, we need to go back to basics. Take London offices as a convenient example. The developers of the immediate post-war period usually attempted to negotiate a lease as long as possible with the tenant. Inflation was not perceived as an important factor, and leases of 99 years without rent review or with only one review at the halfway point were not unknown.

The developer reckoned to create a value for himself from the difference between his development costs and the value of the building on completion. And, in the days when borrowing costs were close to, or below, the yields on which offices were valued, the developer who chose to hang on to his completed development as an investment looked forward to a continuing income surplus from the difference between the rent he received and the interest he paid on his long-term fixed-rate finance. Subsequent increases in capital value were not at the forefront of his mind.

As inflation gradually asserted its grip, and the shortage of central London offices forced rents up in an expanding economy, the position changed. The property development and investment companies wanted to benefit from the subsequent increase in rents as well as from the initial development surplus. Typical rent review periods came down first to 21 years, then to 14 years, then to seven years and finally to today's usual five years.

Until the 1990s the "upward only" clause built into the rent review, which meant that rents could rise but never fall for the duration of a lease, was not perceived as a great problem by tenants. Rental levels fluctuated a certain amount, but the overall trend was strongly upwards and it was considered very unlikely that rental levels at any point would be lower than they had been five years earlier. It therefore cost tenants very little to concede the "upward only" clause.

A property investor, however, was doubly protected under the typical 25-year "institutional" lease with five-yearly upward-only rent reviews that had evolved by the 1980s. He had the security offered by the property itself and generally reckoned that he would have little problem in re-letting it if the existing tenant went bust. And, in addition, he had the undertaking by the tenant to pay for 25 years a rent that would never fall below the starting level — in practice, of course, the landlord expected it to increase at five-yearly intervals.

This structure had always assisted considerably in property financing in Britain. With a virtually guaranteed income for at least 25 years, it was possible to finance property investment with very long-term borrowings — a 25-year mortgage being typical.

Allied with good rental growth rates, resulting partly from a planning environment that restricted the supply of property, this lease pattern meant that good-quality property in Britain tended to be valued on lower yields than in many other countries. In other words, investors in the UK were prepared to accept a low return from property at the outset because they regarded the growth prospects as high and the income as exceptionally secure.

This was the background. Now look at the changes which took place in the early 1990s. Excessive development combined with general economic recession left many buildings empty. Rental levels began to fall. The balance of power switched from the landlord to the tenant, or at least to the prospective tenant. Rental levels would often be lower at the review point than five years earlier. Only the "upward-only" clause protected the landlord's income. He could have no certainty that he could re-let a building that became empty. If he did succeed, it would probably be at a lower rent.

These factors help to explain the changes in attitude to property investment and the lease structure which took place during the 1990s market slump. The conventional wisdom in the property world used to be that the virtues of property were determined by "location, location and location". Location is still important. But in the early 1990s the covenant — the standing of the tenant and his ability to go on paying the rent — acquired far greater significance for a time. Temporarily at least, property investment became less about property and more about latching on to the cash flow of a strong company.

At the same time prospective tenants, sensing that the strength of their bargaining position had increased, tried to challenge the traditional institutional lease structure. They objected to committing themselves for 25 years and they objected to the "upward only" rent review clause. A 10- or 15-year lease, possibly with a break clause, was more to their taste.

The traditional big property investors, notably the insurance companies, generally tried to resist this pressure, which they saw as eroding the virtues of property investment from their viewpoint. The result, it has been suggested, might be a two-tier structure with a property let for 25 years with upward-only reviews changing hands on, say, a 7% yield and an otherwise identical property let for a shorter period without the upward-only clause commanding a lower rating: say, a yield of 8.5%.

There was some evidence that this was happening in the early 1990s, though many property market observers considered at the time that it was only a temporary phenomenon and that the 25-year lease would become almost universal once again when supply and demand in the market returned to a better balance.

How would you value a building let for 10 years with a break clause after five years, compared with an identical building let for 25 years with upward-only rent reviews? There is a fair consensus of opinion that you would value it on an internal rate of return (IRR) basis rather than by reference to a traditional all-risks yield. The result could, however, be expressed in terms of an all-risks yield.

The important point to remember about IRR calculations is that you need to specify your assumptions. This is very different from the traditional all-risks yield approach where, the cynics would have you believe, the valuer simply concludes that "9% (or whatever) feels about right" for a particular building.

So let us take an example. For purposes of illustration we will concentrate on the main elements in the valuation, ignoring items such as transaction and letting costs, which in practice would need to be taken into account. And we will adopt the usual convention that rents are received annually in arrear (whereas in reality they are received quarterly in advance).

Suppose the building in question is a freehold office block producing a rent of £350,000 a year. Suppose, too, that if it were let for 25 years to a good tenant on a typical institutional lease, it would be valued on a 9% yield. This means it would be valued at 11.11 "years purchase"). On this basis it would be worth £3.89m (£350,000 multiplied by 11.11).

Though they are not specified, we might guess at some of the assumptions that probably underly this valuation. An investor would be looking, say, for a return about 1.5 to 2 percentage points above the absolutely safe yield he could get by buying a government bond ("gilt-edged stock"). And he would perhaps assume that the rental value of the building would grow at a compound rate between 3% and 4% a year in the long run, though would possibly be expecting very little if any growth in the immediate future. If he was right in this assumption over 25 years, a discounted cash flow calculation shows that his IRR — the true return on his outlay — would be around 11.5%. Which might be about right against a gilt yield of 9.5% or so.

Now take the same building, again let at £350,000 a year but let on a 10-year lease with a break clause after five years. In the market of the early 1990s the valuer would probably have assumed the worst and reckoned that the tenant would leave at the earliest opportunity. He would also assume that it might take the owner one to two years to relet the building. He would also probably assume fairly substantial costs to relet the building, but we will ignore these for simplicity and plump for a two-year void period.

So, if we undertake an IRR calculation over 22 years, what do we have? There is a rent of £350,000 coming in each year for five years. Then we assume that no rent is received in years six and seven: the void period. By year eight the building is relet. Assume the new lease is for 15 years without a break clause. But, since we are erring on the side of pessimism, assume, that the initial rent is no higher than the £350,000 received at the outset. Perhaps after this we could be a little more optimistic and assume that rents grow by a more normal 6% a year thereafter. The rent the landlord receives under the new lease therefore rises every five years at the review point. At the end of the 15 year lease — which is 22 years from our starting-point — assume the building is sold at a fairly low residual value which reflects the need for substantial outlay on refurbishment.

Then remember the principles of discounted cash flow (DCF) and net present value (NPV) calculation, — see the early chapters for a detailed explanation. Basically, the principle is that you discount back all outlays and receipts for the relevant periods. The first year's rent is discounted for one year, the second year's for two years and so on until year 22 where both the rent and the assumed price on sale will be discounted for 22 years.

The rate of discount you apply to each of the receipts is your target rate of return: suppose it is the same 11.5% that the valuer seemed to be assuming in the earlier example. Discount at 11.5% for the relevant periods, then add together the discounted present values of all the receipts, including the assumed sales proceeds. The answer comes out at £3.33m. This is the price a purchaser could afford to pay for the building — assuming his assumptions are correct — to show him the 11.5% return that we have reckoned he requires. If he requires a 12% return, he could afford to pay only £3.17m, and so on. In practice, because of the greater risk, the few potential buyers of buildings with 5-year break clauses might require rather higher prospective rates of return than this. At a 15% IRR the value of the property would be only £2.45m on our assumptions. But since this would imply a starting yield of 14.5%, the valuer would probably consider that the discount for the short lease was too great and would re-examine his assumptions.

	Value £m	Starting yield %
Let on 25-year lease	3.89	9.0
Let on 10-year lease with 5-year break		
11.5% target IRR	3.33	10.5
12% target IRR	3.17	11.0
15% target IRR	2.42	14.5

Some different results given by different assumptions are shown in the table. Our examples, we emphasise again, are simply to demonstrate the principles. In a bombed-out property market there are few buyers for properties with short leases or break clauses and there is considerable leeway in the assumptions made, which would vary considerably with the particular property and the state of the market. And the DCF calculation might, in practice, be performed over a shorter period. Our calculation over 22 years is to demonstrate the principle.

But the exercise does show how a two-tier market could emerge, with the short-leased properties changing hands on significantly higher yields than those let on the more traditional 25 year lease. A corollary of this is that — in a market that had returned to better balance — tenants would almost certainly have to expect to pay higher rents than they would under a 25-year lease for the greater flexibility of a shorter lease, one with break clauses or a lease without upward-only rent reviews.

52 Who does what in property

We have looked at the financial mechanisms of the property market and to complement the figurework we need to look briefly at the "players". Who performs what functions in the property business?

The answers are not quite as easy as they would have been in the 1960s or 1970s. Traditionally, the focal point of the property world was the West End of London, where most of the major firms of surveyors and estate agents have headquarters offices. There was a wide gap between the estate agency world and the markets and banking activities of the City of London. The property world did not understand the City very well and the City by and large did not understand property.

By the 1990s that had changed (though not quite as much as some practitioners, both in the banking world and the property world, might wish to think). Property was coming to be regarded as an investment like any other. Though some of its characteristics were obviously different from those of government stocks (gilts) or ordinary shares (equities), it was felt that sensible comparisons ought to be possible between the returns on these different forms of investment.

If comparisons were to be made, the banker or broker would need to have some understanding of property and the property expert would need at least a working knowledge of equities, gilts and the money markets of the City. It was thought at first that, as part of the deregulation process in the financial markets, surveyors would perhaps merge with City securities houses so that the client could shop for property expertise and banking or stock market advice in the same establishment.

By and large, this has not yet happened on any large scale. The major firms of surveyors and estate agents remain independent, though one or two medium-sized or smaller ones have been taken over by financial groups. (We are talking here of the agents specialising in commercial and industrial property. The position is very different among residential estate agencies, a large proportion of which were acquired — often at excessive cost — by financial institutions in the latter years of the 1980s).

Instead, the financial groups of the City have tended to try to develop in-house expertise on commercial property, while many of the larger estate agencies set up financial services subsidiaries, sometimes employing former bankers or City men in an attempt to broaden their areas of expertise.

This trend was heightened by the various attempts to "securitise" property: to transmute the physical investment into various forms of paper investment that could be traded in the traditional securities markets. As we have seen, initially at least these attempts met with very limited success. But at the same time, bankers were applying to property financing many of the more innovative "financial engineering" techniques developed initially for the financing of industrial and commercial companies.

A third important change at the end of the 1980s and beginning of the 1990s was that the traditional savings institutions — the insurance companies and the pension funds — had become less important as a source of finance for commercial property and the banking system had become vastly more important. The major firms of surveyors had traditionally acted as intermediaries between property developers and institutional sources of finance.

In combination, these trends threatened to deprive the major estate agents and surveyors of some of their most profitable business, which they saw disappearing towards the securities and banking markets of the City.

It is against this background that our comments on the main players in the property market need to be read.

Surveyors and estate agents. This must be the starting point for any discussion of the commercial property market in Britain. We are not too concerned with the technical distinction between firms of surveyors and estate agents. The businesses we are referring to are the firms or companies which fulfill an agency role in the commercial property market.

The major agencies undertake a vast range of activities, but we are concerned mainly with those touching on the financial and investment aspects of property. First, in a professional capacity surveyors value properties. Since there is no central market as in shares, and since no two properties are alike, this is a highly important function. We have seen that professional valuations are likely to be needed in stock market launches, takeovers, annual accounts, for setting the price of property bonds and in many other contexts.

The valuation function merges into the advice function. The major agencies advise buyers and sellers on individual transactions, they may be retained as advisers to major property-owning funds or may even manage these funds on behalf of the owners. They are also likely to be called on to advise on developments, to evaluate development plans (sometimes on behalf of banking lenders). Not infrequently, they manage a development on behalf of its promoters. Advice extends to negotiating leases and rent reviews for property owners or tenants and management functions can include the physical management of buildings on behalf of the owners.

Estate agents put property buyers in touch with sellers, and *vice versa*. They put investors in touch with entrepreneurs seeking finance. They put tenants in touch with landlords. From the vast number of transactions passing through their hands they build up a feel for the way the property market and its many sub-markets are moving. This information is one of their most valuable assets, given the lack of a central and visible market for property.

The agents' knowledge of the market is extended and codified by the research operations the major agents run. Research embraces trends in the rental markets and investment markets (sometimes published in the form of widely-available indexes) and research into supply and demand for particular types of property in particular regions. The major agents will also undertake one-off research for clients considering major developments, particularly in the retail sector: catchment areas, spending patterns and the like.

We have already mentioned the financial services arms of the major agents. In essence, the more active ones attempted to bridge the gap between the property and the financial markets. Each had its own area of specialisation, but their activities included: advising private (and occasionally public) companies on financial structuring; helping entrepreneurs to decide on financing methods, helping them to prepare proposals in a way acceptable to lenders and investors and putting them in touch with suitable sources of finance; researching new methods of financing property (this comes under the general heading of "financial engineering", where the agents are competing head-on with the banks) and even setting up property funds; promoting various schemes for unitisation of property (this was seen as a major source of profit if unitisation ever got off the ground).

Property research organisations. Independent of the major agencies, there has been a proliferation in recent years of

specialist research "boutiques" which will research and advise on most aspects of property investment and development. They are not the same thing as the property finance "boutiques" (see below).

Auctioneers. The main property auctioneers are also estate agents. While most major investment property sales are negotiated "by private treaty", the auction method is also used and is particularly attractive for secondary and tertiary property and for smaller lots in which private investors as well as companies and institutions may be interested. Auctions are the most visible manifestation of the property market at work.

Accountants. Company accounts, prospectuses, and takeover documents all require attention from the accountant. The accountancy firms fulfil a more active role in advising on the tax aspects of property ownership and development — a very important aspect of most property planning.

Solicitors. Few property transactions can take place without lawyers becoming involved: conveyancing of properties, legal aspects of leases, litigation and so on. Some of the more specialist firms of solicitors take an active role in devising new forms of property ownership and work with merchant banks and accountants on the structural aspects of various forms of financial engineering associated with property.

Stockbrokers. Launches on the stock market require a sponsor who is frequently a broker. Stockbrokers' analysts also undertake a great deal of research into property companies and, by extension, follow the direct property market closely.

Banks. The merchant banks and some of the US investment banks study property quite closely and advise on financing structures. Their role is discussed in more detail below.

The planners. Decisions by the planning departments of local authorities can be worth many millions of pounds to prospective developers. So can decisions by the Department of the Environment, which adjudicates on appeals against local authority decisions.

Insurance companies and pension funds. Insurance companies and pension funds were the traditional long-term owners of commercial property investments. The life assurance funds and the pension funds both have control of long-term investment money for which property provided a suitable home. Since both were investing their own (or their savers') money, rather than relying on borrowed funds, capital appreciation was generally as important to them as income and they were not troubled by the income deficit problems of property owners who rely on borrowed money.

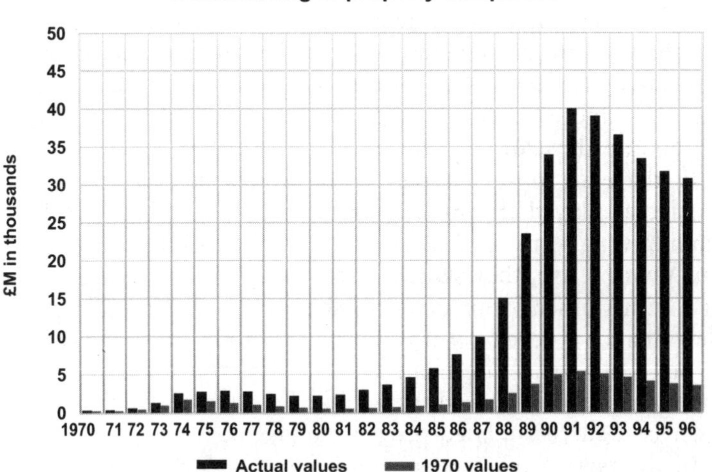

Whether you look at the actual amounts or the inflation-adjusted figures, bank lending for commercial property provides the clearest possible warning of coming trouble in the property sector. The property market collapses of the mid-1970s and of the early 1990s were both preceded by a massive escalation of bank lending for property. Some cynical property investors with good memories tend to look at what the banks are doing and do the opposite: the aftermath of a crash, when the banks are trying to reduce their outstanding property loans, is often the time when investors should be buying property. Source: *Bank of England; DTZ Debenham Thorpe*

The bigger insurance companies and the very large pension funds have extensive property departments, capable in many cases of undertaking developments in their own right as well as buying and managing completed investment properties. Smaller pension funds may invest indirectly by way of the unauthorised property unit trusts.

However, the excellent performance of equities for much of the 1980s and a rather poor performance by property until the closing years of the decade tended to detract from the appeal of property. And, needless to say, the property market crash of the early 1990s has not helped. The insurance companies, with a longer history in the business, were less affected than the pension funds which — except for some of the very large players — severely cut back their net new investment in property. However, both types of institution dealt more actively than in the past as they rearranged their portfolios; gross purchases were considerably larger than the net additions. Official figures for 1995 showed UK property holdings of the insurance companies (life and general combined) at £39bn and those of the pension funds at £24bn.

In the early post war period the insurance companies in particular played an important role in property financing by lending long-term fixed-interest funds. When inflation made this unattractive, they switched to providing leaseback funds or various other forms of profit-share finance. Nowadays, if they finance a developer it is more likely to be via some form of pre-purchase arrangement whereby the property ends up in the ownership of the institution and the developer effectively receives a success-related project management fee.

In the past, bank lenders for property development have relied on the institutions to buy the completed projects as long-term investments, thus enabling them to get their money back. In the 1990s it is doubtful if the institutions will provide a long-term "take-out" for more than a small proportion of the property the banks are currently financing.

Property companies. Publicly-listed and private property companies were a significant force in the market in the second half of the 1980s, buying both for development and for investment. At the end of the decade the total property assets of listed property companies were reckoned to be in the region of £40bn. After the slump the figure was probably closer to £30bn.

As we have seen earlier, the commercial property company market breaks down broadly between investment companies, traders and dealers, though with considerable blurring of the boundaries. Investment companies and traders normally undertake developments, the former holding on to the completed property as an investment and the latter selling it for a trading profit. The dealers are those who simply buy properties or land with a view to selling at a profit.

Because of the deficit-financing problems of funding long-term property investments on borrowed money, most of the newer property companies coming to the stock market in the 1980s were initially traders or dealers. The large investment companies generally have their origins in an earlier era. The newcomers financed their developments by extensive use of bank money, often in off balance sheet companies. And many of them failed to survive the 1990-93 mayhem in the property market. By contrast, the revival of property values in late 1993 and early 1994 bought a number of new asset-based property companies to the market early in 1994.

Overseas buyers. As British institutions became more wary of property in the late 1980s, overseas buyers became a more important factor in the British market. Japanese, Scandinavian and American investors (and in some cases, developers) were making substantial purchases towards the end of the decade, often concentrating on the so-called "landmark" buildings in Central London: the absolutely prime sites in the main business areas. Many of them were severely burned during the slump. Others, notably Germans, came in search of bargains once values had fallen.

Private individuals. With lower tax rates and easier accumulation of wealth in the 1980s, rich private individuals from Britain and overseas have become a more important factor in the market, buying a number of significant properties as well as the more usual smaller auction lots. But the distinction between the individual and the company is blurred. Many individual entrepreneurs and investors tend to operate through a company vehicle, whether listed or not.

Construction companies. The major contractors are important to the developers: they undertake the physical construction of the buildings. But many construction companies also undertake development on their own account, either holding on to the completed properties or selling them for a trading profit on completion. Some of the major names are or have been significant owners of investment property. Occasionally they will also act as a source of construction finance for the developer.

Industrial and commercial companies. Many companies own the premises they operate from: retailers in particular. At the

end of the 1980s the trend was to try to finance the properties off balance sheet, retaining an interest in future growth in values but releasing funds to employ in the core business. In the new accounting climate of the 1990s, the property and the associated debt are more likely to show up in the accounts.

Who finances property?

By the end of the 1980s, **bank** lending to property companies had reached an all-time record of over £30bn, having escalated very rapidly over the previous few years, and the Governor of the Bank of England was warning the banking system — not for the first time — to rein back. The banks failed to heed the warning in time. Bank lending for property actually peaked at more than £40bn in 1991, but by then Britain was well into the recession.

This lending was for a mix of development and investment projects. The development loans were, by their nature, short-term: typically, two or three years, or longer for very large projects. The normal pattern was for a development loan to be repaid from the proceeds of selling the development once it was completed and let. But with a sticky investment market at the end of the decade, many developers were trying to transmute their development loans into investment loans for five years or more, reckoning to hold on to the property until the first rent review.

The clearing banks are major providers of corporate finance facilities to property companies, as well lending on many smaller projects through their branch networks. Major overseas banks are active in syndicated project finance loans, often made on limited-recourse terms to development companies. For some overseas banks, property loans constituted a very large proportion of their lending in Britain at the end of the 1980s. Most of them subsequently drew in their horns. In the 1990s some German property banks emerged as significant players in the property investment loan market.

In project finance, a general distinction needs to be made between the lenders who are happiest simply putting up senior debt — the lowest-risk part of the lending operation — and those which are prepared to take more risk in return for higher interest rates or a profit share, though sometimes one bank will fulfill both roles.

The mezzanine/profit-share lending operation is mainly the province of merchant banks, some overseas banks and a few specialist lending institutions. It requires more property expertise than simply providing senior debt. In addition to **specialist property lenders** there are a number of **property finance advisers** which structure financing packages in return for a fee or a commission, without providing the funds themselves.

One or two of the American banks and investment houses have developed a very strong in-house property capability which allows them, as well as providing finance, to market major buildings and occasionally whole property companies. In this respect they sometimes seemed ready to challenge the traditional British surveying firms.

In the late 1980s some of the larger **building societies** were beginning to compete at the fringe with the banks in financing certain commercial developments. Their subsequent experience was not always encouraging, though some remained committed to the commercial market through the slump and beyond.

Finally (and perhaps briefly), in the late 1980s **insurers and insurance brokers** came to be counted as part of the property finance armoury. By the turn of the decade it was quite common for certain lenders to require insurance of either the top slice of the loan they advanced, or sometimes the whole amount. The top slice insurance is straightforward risk insurance. Insurance of the whole loan ("bottom up" insurance) is more a matter of credit enhancement, making it easier to syndicate the debt with the covenant of major insurance groups behind it.

Specialist insurance brokers structure financing packages incorporating various levels of insurance for the loans, while the risk is underwritten by a major insurance company which probably itself spreads the risk via syndication or reinsurance.

However, the collapse in property values in 1990-93 resulted in many claims – and frequent disputed claims – under these loan insurance packages and involved large losses for some insurance providers. There was a general withdrawal from this type of business.

The very rapid growth in bank lending for property up to the end of the 1980s inevitably meant that many of those involved had comparatively little experience of the property world and had certainly never undergone a severe bear market such as occurred in the period 1974-76. This inexperience was often reflected in the lending and borrowing decisions of the time, which frequently verged on the super-optimistic. By the early 1990s the chickens had come home to roost. But it was an expensive learning process both for banking lenders and for developers.

Perhaps the full extent of the 1990-93 property market crash could not have been foreseen. Perhaps its lessons will be taken to heart. But even if we ignore cataclysmic events of this nature, it remains to be seen whether the banking world has really bridged the expertise gap between the City and the West End world of the property men and women — and whether the property industry is yet completely comfortable in its dealings with the financial markets.

Glossary

amortise, depreciate, To amortise or depreciate an asset is to write it off in accounting terms over its life or estimated life. Thus a leasehold interest needs to be written down to a nil value by the time the lease expires. Physical assets which wear out also need to be depreciated. Unfortunately, the amounts set aside from profits to amortise a lease or the value of a property are not (except in the case of industrial property) allowed as a deduction for tax purposes. Tax relief can, however, be obtained on depreciation of the plant and equipment that a building contains.

auditors, audit, Auditors are independent accountants who verify the annual accounts prepared by the management of a company and certify to the shareholders in the auditors' report that they present a "true and fair view" (or, if they do not, point out why not). Anyone examining the published accounts of a company should check on the auditors' report to see if there are any qualifications of significance. (see also qualifications, going concern basis)

base rate, bank base rate, The interest rate yardstick for traditional bank overdrafts and loans, rather than for the money market loans which larger companies will tend to tap. Base rates are changed periodically to reflect the cost of money in the economy or to reflect the government's interest rate intentions. Borrowers pay a margin over base rate and depositors get a rate of interest below base rate.

basis point, A basis point is one hundredth of a percentage point. Thus 150 basis points is the same as 1.5% or 1.5 percentage points. The cost of borrowings in the money markets, where margins can be very slim, is usually expressed as "so many basis points" over or under LIBOR (qv). On the other hand, interest rates on the more traditional overdraft will probably be expressed as "so many points" (meaning percentage points) over bank base rates (qv)

bear, bearish, A bear expects prices (particularly in the stock market) to fall. Hence a "bear market" is used to describe a lengthy period of falling prices. A bull (qv) is the opposite of a bear.

break-up net asset value, Sometimes used to mean the net asset value of a company after deduction of the tax that would be payable if the company's properties were, in fact, sold.

bull, bullish, A bull is someone who expects prices to rise, particularly in the stock market but also sometimes in the property market. Hence a "bull market" is a market where prices are on a rising trend and somebody who is "bullish" about the property market would be expecting values to rise. A bear (qv) is the opposite of a bull.

cap, Upper limit on interest rate, etc

capitalise, capitalisation, To capitalise interest is to add it to the cost of a project rather than charging it against revenue. To capitalise an income flow is to put a present capital value on it; thus, if an investor is prepared to buy a property at 15 times the rent it produces (at 15 years' purchase) then 15 is the capitalisation rate. The equity market capitalisation of a company is the total value of all the ordinary shares in the company at the current market price.

cash flow, The key to property finance. Companies go bust because they run out of money - they do not have the cash to meet their interest payments and other obligations. Cash flow is not the same as profit. A home owner may have bought his house at £50,000 and find it it now worth £60,000 (on paper he has made a profit of £10,000), but if he does not have the income to cover his mortgage payments he may lose his home. The same phenomenon occurs on a larger scale with property companies. A positive cash flow is one where cash coming in exceeds cash going out. This may be applied simply to income (revenue receipts exceed revenue outflows) or it may be applied to the overall position of the company - what it is getting in (including receipts from sales of properties) exceeds the cash flowing out (for salaries, interest payments and capital expenditure on developments).

conversion premium, In the case of a convertible stock, the difference between the conversion price and the company's share price in the market when the stock is announced. If Payola Properties' shares stood at 100p and it issued a convertible stock which converted at 120p, the conversion premium would be 20%.

convertible mortgages, There is much talk of the development of convertible mortgages, where the secured lender receives a return geared in some way to the income from or value of the property. These exist in some other countries but, as explained in the text, pose problems in the UK for tax and legal reasons. The attraction is that they would carry a lower interest rate initially in return for what is effectively a profit share.

covenant, This technically means an undertaking given by a tenant or a borrower. A restrictive covenant in a lease might prevent a tenant from sub-letting a property and a covenant in a loan agreement might limit the borrower's total debt according to some formula. However, by extension the word "covenant" in property language often refers to the standing of (and therefore the security offered by) the tenant. Unilever or ICI would be an excellent covenant. (See also restrictive covenants)

Datastream, An on-line database of company, share price and economic information, from which the graphs in this book have been prepared. The service includes the property performance indexes prepared by surveyors Hillier Parker.

deep discount bond, A bond which offers at least part of its return as a gain to redemption. The extreme example is a zero-coupon bond which pays no interest at all. If it were issued at £50 and repaid at £100 after seven years, it would offer a return equivalent to about 10.4% a year compound over its life. Zero-coupon bonds can be useful in property financing, where income is short in the early years.

deep discount rights issue, A rights issue of shares made at a price a long way below the current market price of the shares: say, at a discount of 50% rather than the 20% or so of the conventional rights issue. A deep-discount rights issue will probably not be underwritten.

derivatives: There have been various attempts to "manufacture" property derivative products in the 1990s. Derivatives are "synthetic" financial products whose performance reflects the performance of underlying assets. The idea is that these derivatives would allow investors to take a bet on property performance – or protect themselves against adverse movements in property values – without needing to buy or sell physical properties. Early in the decade the commodity market then known as London FOX introduced futures contracts in commercial rents and values. Participants could back their view of future movements in the commercial property market by buying or selling property futures, whose value was linked to movements in IPD indexes. Or that was the theory. In the event, the initiative collapsed in a major scandal at London FOX (which in no way reflected on the integrity of IPD). In the second half of the decade there were more soundly-based moves by a group of heavyweight financial institutions to establish a traded market in property derivatives, to be know as the Real Estate Investment Market or REIM. However, the project encountered considerable delays. In the interim, BZW – the investment banking arm of Barclays Bank – had introduced over-the-counter derivatives in the form of Property Index Forwards. Like futures, these forward contracts allowed investors to back their view of property performance, again measured by reference to the IPD index.

dividend cover, The number of times a dividend payment is covered by the available profit. It is thus a measure of the safety of the dividend. In principle, if a company earns net profits of £6m and the net dividend absorbs £2m, the dividend cover will be three times. In practice, tax factors may make the calculation a little more complex if the company has significant overseas earnings.

equity, In stockmarket and property parlance, equity is used to mean the interest in a company or a property which bears the full risk and receives the full rewards with no upper limit. It is the entrepreneur's or owner's money, as opposed to borrowed money. Thus "equities" on the stock market are simply ordinary shares. Preference shares represent owners' money, but are not equity because they have preferential rights and a fixed dividend. The equity funds of a company are the ordinary shares plus the reserves, which represent the ordinary shareholder's total interest in accounting terms. If you are the equity owner of a property, you own it outright. If you buy a £1m property with £700,000 of borrowed money and £300,000 of your own, your equity interest in the property is £300,000.

going concern basis (see also auditors), If the auditors to a company, in their report, need to point out that the accounts have been prepared on a going concern basis, it usually means that they have doubts whether the company is, in fact, financially viable. Accounts are normally prepared on the basis that a company will be continuing in business: a going concern. If it is insolvent, accounts would be prepared on a break-up basis. In this case assets would probably be stated at the considerably lower figures they might achieve in a forced sale.

gross, net, Gross usually means before deductions and net means after deductions (particularly tax). Gross rents are rents before ground rents and management charges have been deducted, net rental income is what remains after deducting (particularly) ground rents. Net dividends are after deduction of tax; gross dividends are the equivalent item before tax. To "gross up" a net figure is to calculate the equivalent before the tax deduction.

industrial property, Usually refers to factories and warehouses, which traditionally offer a higher yield than shops and offices (see also sheds).

insurance (of a property loan), There are two main types of property loan insurance. With top slice insurance, the insurer provides a guarantee for the top part of the loan. A banker might normally be prepared to advance 70% of value, but would perhaps increase this to 85% if an insurer took the risk on the top 15%. With ground-up insurance, the whole of the loan is insured against loss to the lender. As this is less risky, the premium rates are obviously lower. The main purpose of ground-up insurance is to substitute the backing of an insurance company for the security offered by the property project itself. This may make the loan a lot easier to syndicate among lenders who have no great experience of evaluating property projects.

Investment Property Databank (IPD), is an organisation that collates, monitors and analyses the property holdings of major British investors. Its database covers some £55bn of commercial property in the UK, representing over 90% of the investing institutions' commercial property holdings plus a somewhat smaller proportion of property company portfolios. From the information supplied by these commercial property owners, IPD is able to analyse most aspects of commercial property performance in Britain and it also produces indexes covering the different aspects of performance. Its indexes are widely accepted as benchmarks and are used for various types of property derivative product.

junk bonds, A term which originated in the United States to describe corporate bonds that are considered to be below institutional investment grade. To reflect higher risk they will pay considerably higher interest than top grade bonds. Junk bonds arise particularly in the course of leveraged buyouts or takeovers, when debt is substituted for much of the existing equity and the junior layers of debt will carry above-average risk. But some junk bonds are simply bonds issued by companies too small for their securities to be considered of investment grade. Junk bonds are a form of securitized mezzanine debt, and while mezzanine is not uncommon in Britain, its traded equivalent is rare. See also pay in kind securities

lease, full repairing and insuring lease, Institutional investors usually favour buildings let to a single tenant, with the tenant taking full responsibility for fitting out, maintaining and repairing the building. This would be known as a "full repairing and insuring lease" **(FRI)**. A typical "institutional lease" (a lease of the kind an institution would grant) would run for 25 years with upward-only rent reviews every five years.

leveraged takeover, buy-out, Leveraged transactions are ones involving large amounts of debt. A leveraged takeover occurs when a company is bought mainly with borrowed money. Most buy-outs, in particular, involve considerable amounts of leverage. It would not be unusual, in a leveraged bid for a company valued at £100m, for £80m of the purchase price and possibly more to be raised as borrowed money. After the takeover, interest charges and capital repayments absorb most of the company's cash flow in the early years. The high gearing implies considerable risk, and a number of highly-leveraged deals have gone badly wrong as a result, both in the United States and in Britain.

management buy-out, buy-in, A management buy-out occurs when the managers of a business decide to become owners as well, and buy the business they run from its existing proprietors. It may be that they buy the subsidiary they run from the parent company. Or the managers of a listed company may launch a takeover offer for the whole business. In either case they will usually need to borrow large amounts of money to make the purchase. A buy-in is where a new management team attempts to take over a company that it does not at present run. (see also leveraged takeover, buy-out)

margin, spread, "Margin" or "spread" are terms often used to describe interest rate differentials relative to some accepted yardstick. Thus a loan might be granted at a margin of 100 basis points over LIBOR (with LIBOR at 10%, the interest rate would be 11%). Or a company bond might be priced at a spread of 150 basis points over the benchmark gilt-edged stock. If the gilt-edged stock offered a redemption yield of 12%, the company bond would thus be priced to yield 13.5%.

maturity, Refers to the life of a security, usually a bond. If a bond has four years to go to maturity, it will be redeemed (repaid) in four years.

nominal value, par value, face value, In Britain, most forms of security have a nominal or par value. Shares are normally issued as units of 5p, 10p, 20p, 25p, 50p or £1 par value. Bonds in Britain are usually taken to have a par value of £100. The par value of a share is important for some accounting purposes, but has very little to do with the price at which it is traded, which is determined by buyers and sellers in the market. The same applies up to a point with a bond, except that it will normally be repaid at the end of the day at its par value of £100.

open offer, A company issuing new shares for cash to investors who are not already shareholders (perhaps in the course of a vendor placing) may be obliged under pre-emption guidelines to give existing shareholders first bite at the issue. Thus the shares are provisionally placed (sold) to a range of investors, ensuring that the company gets its money, but an open offer is made to existing shareholders, giving them the right to "claw back" these shares at the same price if they want them. The investors with whom the shares were provisionally placed thus keep only those shares which are not wanted by existing shareholders. A "placing and open offer" thus sometimes replaces the more traditional rights issue.

pay in kind securities (PIK), These are generally a form of junk bond and, like junk, are more common in the United States than in Britain. Instead of receiving interest each year, investors receive a further issue of the junk bond and the whole issue is redeemed at some future point. The effect is therefore much like that of a zero-coupon bond.

point, percentage point, If bank base rates are 12% and an overdraft costs three points over base rate, the reference will be to percentage points. The current cost of the overdraft is thus 15%. (see also basis point)

pre-emption rights system, The system, enshrined in company law, under which new shares offered for sale for cash must first be offered to the existing shareholders of the company - as in a rights issue. Institutional shareholders are very keen on sustaining this system, though they will allow companies to make small issues where the shares are not offered first to existing holders. (see also placing, open offer)

qualifications (in auditors' report accompanying company accounts), If the auditors "qualify" their report, it may be serious or it may not. If the qualification is simply a minor one - the accounts differ in some relatively insignificant way from recommended accountancy practice, for example - it need not cause much concern. If the report points out that proper books of account have not been kept, or that the accounts do not present a true and fair view, the qualification is very important indeed.

real, real return, "Real" is often used to mean "adjusted for inflation" in the context of investment returns or interest rates. If interest rates are 12% but inflation is running at 8%, real interest rates would be around 4%.

retail warehouses, The large, usually edge-of-town hangar-type buildings used by, among others, furniture, do-it-yourself and electrical retailers. Large car parks for customers are normally provided.

reverse premium, Normally, the buyer of a leasehold interest will pay a capital sum - a "premium" to the vendor.

However, should the rent on the building be above market rates, the leaseholder might need to pay the purchaser ("assignee") a lump sum to persuade him to take on this liability. "Reverse premium" is the term often used to describe such a payment.

reverse yield gap, In a property context, this is normally taken to refer to the difference between long-term borrowing rates and the yields on property of a particular class. If a property company debenture cost 11% in interest and a property yielded 6%, the reverse yield gap would be 5%. The word "reverse" crops up because, in earlier non-inflationary days, property yields would have been above long-term borrowing costs; the reverse is now the case.

sheds, Popular term for simple, standard industrial buildings which consist of little more than four walls and a roof. By extension, sometimes applied to retail warehouses (qv).

stag, A speculator who subscribes for shares in a new issue on the stock market in the hope they can be sold for an immediate profit when trading begins. He is not interested in the shares as a long-term investment.

stepped-interest bond, A bond which starts life paying a fairly low rate of interest, which then rises at predetermined points. It might, say, pay 9% for the first five years, rising to 12% at year six and 18% at year 11. This can be useful in property financing, as the increases in interest rate can be designed to coincide with rent reviews on the property, when rising rents will cover the higher coupon. Provided the interest rate increases are stipulated in advance, the bond will not fall foul of tax provisions that would deny tax relief to a bond whose return was geared to profits. However, in accounting terms, the issuing company has to provide for the interest each year at the true average rate over the life of the bond.

underwrite, underwriting, underwriter, Most issues of shares for cash - such as new issues on the stock market and rights issues - are underwritten to ensure that the issuer gets its money. In return for a fee, the underwriter (usually a bank or securities house) agrees to buy all the shares at the issue price if other buyers do not emerge. In turn, the underwriter normally lays off all or part of its risk with sub-underwriters (mainly the big institutional investors) each of which agrees to take up a certain proportion of the issue if required to do so. Typically, underwriting fees might be 2% of the value of the issue, with the underwriter paying out of this 1.25% to the sub-underwriters. Bond issues and syndicated loans may also be underwritten in the sense that the arranging bank agrees to take up the whole amount initially, thus assuming the risk of finding investors or other lenders to which to distribute it. (see also open offer).

valuation tables, The compound interest tables a valuer traditionally uses as an aid to property valuation (which is effectively a matter of calculating the likely present value of a future income flow). They are likely to include leasehold valuation tables assuming different tax rates and sinking funds compounding at different interest rates. Traditionally they worked on the assumption of rents paid annually in arrear but nowadays may include a "quarterly in advance" alternative. Increasingly, they are being supplanted by programmable calculators or computers which can produce the same figures.

vendor placing, This arises when a company making an acquisition wants to pay in shares, but the vendors want cash. The purchaser issues its shares to the vendor but arranges at the same time for these shares to be placed for cash with investors. Thus the vendor receives the cash. The arrangement may include an open offer (qv), giving existing shareholders in the purchaser the chance to claw back the shares from the investors with whom they were placed.

winding up, liquidation When a company is wound up or liquidated, its assets are sold and the money is used to repay creditors in order of precedence. Whatever remains after all debts and preference capital have been repaid belongs to the ordinary shareholders. In practice, if a company is compulsorily liquidated because it gets into financial trouble, there is rarely anything left for the ordinary shareholders and creditors may lose money as well. However, a solvent company may be put into voluntary liquidation, with the intention of returning funds to shareholders. (See also receivership, Insolvency Act 1986)

withholding tax, A tax deducted by a company or borrower and handed over to the revenue authorities before remitting interest or dividends to investors. Tax authorities like withholding taxes because they are sure of receiving the tax, rather than having to rely on the recipient of the interest or dividend to pay it. Investors, particularly in the euromarkets, hate them for obvious reasons.

Company taxation—Stop press

After the main body of this edition of Property and Money had been prepared for press, the first Budget from the new Labour government in July 1997 significantly restructured the system of company taxation in Britain. This has inevitably outdated some of the taxation examples provided in the text.

The main change introduced in the July 1997 Budget was the abolition of the dividend tax credit for pension fund shareholders, who are the stock market's biggest investors. As we had suggested, the tax regime as it existed prior to July 1997 turned out to be merely an interim stage in the evolution of the company tax system. The changes introduced by the new government have a severe adverse impact on pension funds' dividend income and significantly increase the disadvantages for pension funds of holding property investments via a company structure. On the other hand, rental income from property investments held direct by the pension funds continues to come through without deduction of tax.

At the same time as abolishing the pension funds' right to reclaim the dividend tax credit, the new government reduced the rate of corporation tax from 33% to 31%.

Prior to the July 1997 Budget, companies (including property companies) were obliged to pay Advance Corporation Tax or ACT on the dividends they distributed. ACT was charged at a rate of 20%. Thus, a company distributing a net dividend of 8p to shareholders had to pay 2p of ACT to the Inland Revenue, since 10p before tax is the amount that provides 8p after tax of 20%.

Shareholders who were not liable to tax, like the pension funds, received a tax credit with their 8p net dividend which allowed them to claim back the 2p (in our example) of tax that had been paid. An 8p net dividend was thus worth 10p to the pension funds. Now that they may no longer reclaim the tax credit, an 8p dividend is worth just 8p.

Companies were (and remain) able to offset the amounts they have paid in ACT against their main corporation tax charge. To see the effects of the Budget changes on companies and their shareholders, look at the following example. To emphasise the point, we have assumed a company that distributes the whole of its income as dividends and whose shareholders are all pension funds:

	Before July 1997 Budget changes		After July 1997 Budget changes	
	£	£	£	£
Pre-tax profit:		100		100
Corporation tax				
ACT on dividends:	16.75		17.25	
Mainstream Corporation tax:	16.25		13.75	
		33		31
Net profit after tax:		67		69
Distributed as dividends:		67		69
Tax reclaimable by shareholders:		16.75		nil
Shareholders' total receipts:		83.75		69

Thus, for every £100 of pre-tax rental income received by a property company, the maximum available for distribution to shareholders rises slightly from £67 to £69 thanks to the reduction from 33% to 31% in the corporation tax rate. But for a pension fund shareholder this benefit is far outweighed by the fact that it can no longer claim the dividend tax credit. The combined effect of the changes is that the maximum amount a pension fund shareholder could receive from every £100 of rental income earned by a property company reduces from £83.75 before the budget to £69 afterwards. Tax-paying private shareholders are in a rather better position. The dividend tax credit still covers their liability to basic rate income tax on dividends received.

A month or so after the Budget it is still too early to say what effect the tax changes will have on property companies' dividend policies. Some may decide to distribute a smaller proportion of income in future, since the total tax burden on distributed profits has risen so sharply for pension fund shareholders. However, the benefits for pension funds of holding property investments direct rather than through a company structure has been very sharply underlined. And since dividends from industrial and commercial companies also become worth less in the hand of pension fund shareholders, the merits of direct property investment relative to investment in equities may have been enhanced. The table below shows the maximum that pension fund shareholders could receive from income deriving from different sources, before and after the July 1997 Budget:

Maximum amount receivable by a pension fund from £100 of pre-tax profits, rents or interest before and after the July 1997 Budget

	From investment in			
Maximum amount receivable by pension fund	Industrial and commercial companies £	Property companies £	Direct property holdings £	Fixed-interest bonds £
Before Budget	83.75	83.75	100	100
After Budget	69.00	69.00	100	100

The Budget changes also complicate the question of dividend yields. Previously, dividend yields on shares had normally been expressed gross (before tax). Thus, in our example of a company that paid a net dividend of 8p that was worth 10p to a pension fund shareholder, the dividend yield would have been calculated on the 10p gross amount. Following the Budget changes it seems likely that dividend yields will be calculated on the net dividend rather than the gross amount. This, of course, has the effect of reducing the average level of dividend yield in the stockmarket. To see the impact, look at the effect of the change on a company that paid a net dividend of 8p per share and whose shares stood at, say, 300p in the market:

Budget effect on dividend yield

Before July 1997 Budget changes

$$\frac{\text{Gross dividend 10p}}{\text{Share price 300p}} \times 100 = \text{gross dividend yield 3.3\%}$$

After July 1997 Budget changes

$$\frac{\text{Net dividend 8p}}{\text{Share price 300p}} \times 100 = \text{net dividend yield 2.7\%}$$

Subject Index

Bold page numbers refer to the glossary.